THE CHIP

THE
CHIP

How Two Americans

Invented the Microchip and

Launched a Revolution

T. R. Reid

Random House Trade Paperbacks

New York

This is a revised edition of a work originally published in hardcover
in 1985 by Simon & Schuster, Inc., and in
paperback by Touchstone in 1986. This edition is
published by arrangement with the author.

Library of Congress Cataloging-in-Publication Data

Reid, T. R.
The chip: how two Americans invented the microchip and
launched a revolution / T. R. Reid.
p. cm.
Includes bibliographical references and index.
ISBN 0-375-75828-3
1. Kilby, Jack S., 1923–   2. Noyce, Robert N., 1927–1990
3. Microelectronics—History. I. Title.

TK7874 .R43 2001
621.381'0973—dc21
2001019694

Random House website address: www.atrandom.com
Printed in the United States of America on acid-free paper
2 4 6 8 9 7 5 3

Matri filiisque in amore dedicatus.

CONTENTS

THE CHIP

THE MONOLITHIC IDEA

The idea occurred to Jack Kilby at the height of summer, when everyone else was on vacation and he had the lab to himself. It was an idea, as events would prove, of literally cosmic dimensions, an idea that would be honored in the textbooks with a name of its own: the monolithic idea. The idea would eventually win Kilby the Nobel Prize in Physics. This was slightly anomalous, because Jack had no training whatsoever in physics; the Royal Swedish Academy of Sciences was willing to overlook that minor detail because Jack's idea did, after all, change the daily life of almost everyone on earth for the better. But all that was in the future. At the time Kilby hit on the monolithic idea—it was July 1958—he only hoped that his boss would let him build a model and give the new idea a try.

The boss was still an unknown quantity. It had been less than two months since Jack Kilby arrived in Dallas to begin work at Texas Instruments, and the new employee did not yet have a firm sense of where he stood. Jack had been delighted and flattered when Willis Adcock, the famous silicon pioneer, had offered him a job at TI's semiconductor research group. It was just about the first lucky break of Jack Kilby's career; he would be working for one of the most prominent firms in electronics, with the kind of colleagues and facilities that could help a hard-working young engineer solve important problems. Still, the pleasure was tempered with some misgivings. Jack's wife, Barbara, and their two young daughters had been happy in Milwaukee, and Jack's career had blossomed there. In a decade working at a small electronics firm called Centralab, Kilby had made twelve patentable inventions (including the reduced titanate capacitor and the steatite-packaged transistor). Each patent brought a small financial bonus from the firm and a huge feeling of satisfaction. Indeed, Jack said later that the most important discovery he made at Centralab was the sheer joy of inventing. It was problem solving, really: you identified the problem, worked through 5 or 50 or 500 possible approaches, found ways to circumvent the limits that nature had built into materials and forces, and perfected the one solution that worked. It was an intense, creative process, and Jack loved it with a passion. It was that infatuation with problem solving that had lured him, at the age of thirty-four, to take a chance on the new job in Dallas. Texas Instruments was an important company, and it was putting him to work on the most important problem in electronics.

By the late 1950s, the problem—the technical journals called it "the interconnections problem" or "the numbers barrier" or, more poetically, "the tyranny of numbers"—was a familiar one to the physicists and engineers who made up the electronics community. But it was still a secret to the rest of the world. In the 1950s, before Chernobyl, before the *Challenger* rocket blew up, before the advent of Internet porn or cell phones that ring in the middle of the opera, the notion of "technological progress" still had only positive connotations. Americans were looking ahead with happy anticipation to a near future when all the creations of science fiction, from Dick Tracy's wrist radio to Buck Rogers's air base on Mars, would become facts of daily life. Already in 1958 you could pull a transistor radio out of your pocket—a radio in your pocket!—and hear news of a giant electronic computer that was receiving signals beamed at the speed of light from a miniaturized transmitter in a man-made satellite orbiting the earth at 18,000 miles per hour. Who could blame people for expecting new miracles tomorrow?

There was an enormous appetite for news about the future, an appetite that magazines and newspapers were happy to feed. The major breakthroughs in biology, genetics, and medicine were still a few years away, but in electronics, the late fifties saw some marvelous innovation almost every month. First came the transistor, the invention that gave birth to the new electronic age—and then there was the tecnetron, the spacistor, the nuvistor, the thyristor. It hardly seemed remarkable when the venerable British journal *New Scientist* predicted the imminent development of a new device, the "neuristor," which would perform all the functions of a human neuron and so make possible the ultimate prosthetic device—the artificial brain. Late in 1956 a *Life* magazine reporter dug out a secret Pentagon plan for a new kind of missile—a troop-carrying missile that could pick up a platoon at a base in the United States and then "loop through outer space and land the troops 500 miles behind enemy lines in less than 30 minutes." A computer in the missile's nose cone would assure the pinpoint accuracy required to make such flights possible. A computer in a nose cone? That was a flight of fancy in itself. The computers of the 1950s were enormous contraptions that filled whole rooms—in some cases, whole buildings—and consumed the power of a locomotive. But that, too, would give way to progress. Sperry-Rand, the maker of UNIVAC, the computer that had leaped to overnight fame on November 4, 1952, when it predicted Dwight Eisenhower's electoral victory one hour after the polls closed, was said to be working on computers that would fit on a desktop. And that would be just the beginning. Soon enough there would be computers in a briefcase, computers in a wristwatch, computers on the head of a pin.

Jack Kilby and his colleagues in the electronics business—the people who were supposed to make all these miracles come true—read the articles with a rueful sense of amusement. There actually were plans on paper to implement just about every fantasy the popular press reported; there were, indeed, preliminary blueprints that went far beyond the popular imagination. Engineers were already making their first rough plans for high-capacity computers that could steer a rocket to the moon or connect every library in the world to a single worldwide web accessible from any desk. But it was all on paper. It was all impossible to produce because of the limitation posed by the tyranny of numbers. The interconnections problem stood as an impassable barrier blocking all future progress in electronics.

And now, on a muggy summer's day in Dallas, Jack Kilby had an idea that might break down the barrier. Right from the start, he thought he might be on to something revolutionary, but he did his best to retain a professional caution. A lot of revolutionary ideas, after all, turn out to have fatal flaws. Day after day, working alone in the empty lab, he went over the idea, scratching pictures in his lab notebook, sketching circuits, planning how he might build a model. As an inventor, Jack knew that a lot of spectacular ideas fall to pieces if you look at them too hard. But this one was different: the more he studied it, the more he looked for flaws, the better it looked.

When his colleagues came back from vacation, Jack showed his notebook to Willis Adcock. "He was enthused," Jack wrote later, "but skeptical." Adcock remembers it the same way. "I was very interested," he recalled afterward. "But what Jack was saying, it was pretty damn cumbersome; you would have had a terrible time trying to produce it." Jack kept pushing for a test of the new idea. But a test would require a model; that could cost $10,000, maybe more. There were other projects around, and Adcock was supposed to move ahead on them.

Jack Kilby is a gentle soul, easygoing and unhurried. A lanky, casual, down-home type with a big leathery face that wraps around an enormous smile, he talks slowly, slowly in a quiet voice that has never lost the soft country twang of Great Bend, Kansas, where he grew up. That deliberate mode of speech reflects a careful, deliberate way of thinking. Adcock, in contrast, is a zesty sprite who talks a mileaminute and still can't keep up with his racing train of thought. That summer, though, it was Kilby who was pushing to race ahead. After all, if they didn't develop this new idea, somebody else might hit on it. Texas Instruments, after all, was hardly the only place in the world where people were trying to overcome the tyranny of numbers.

The monolithic idea occurred to Robert Noyce in the depth of winter—or at least in the mildly chilly season that passes for winter in the sunny valley of San Francisco Bay that is known today, because of that idea, as Silicon Valley. Unlike Kilby, Bob Noyce did not have to check with the boss when he got an idea; at the age of thirty-one, Noyce was the boss.

It was January 1959, and the valley was still largely an agricultural domain, with only a handful of electronics firms sprouting amid the endless peach and prune orchards. One of those pioneering firms, Fairchild Semiconductor, had been started late in 1957 by a group of physicists and engineers who guessed—correctly, as it turned out—that they could become fantastically rich by producing improved versions of transistors and other mechanical devices. The group was long on mechanical talent and short on managerial skills, but one of the founders turned out to have both: Bob Noyce. A slender, square-jawed man who exuded the easy self-assurance of a jet pilot, Noyce had an unbounded curiosity that led him, at one time or another, to take up hobbies ranging from madrigal singing to flying seaplanes. His doctorate was in physics, and his technical specialty was photolithography, an exotic process for printing circuit boards that required state-of-the-art knowledge of photography, chemistry, and circuit design. Like Jack Kilby, Noyce preferred to direct his powerful intelligence at specific problems that needed solving, and he shared with Kilby an intense sense of exhilaration when he found a way to leap over some difficult technical obstacle. At Fairchild, though, he also became fascinated with the discipline of management, and gravitated to the position of director of research and development. In that job, Noyce spent most of his time searching for profitable solutions to the problems facing the electronics industry. In the second half of the 1950s, that meant he was puzzling over things like the optimum alloy to use for base and emitter contacts in double-diffuse transistors, or efficient ways to passivate junctions within the silicon wafer. Those were specific issues involving the precise components Fairchild was producing at the time. But Noyce also gave some thought during the winter of 1958–59 to a much broader concern: the tyranny of numbers.

Unlike the quiet, introverted Kilby, who does his best work alone, thinking carefully through a problem, Noyce was an outgoing, loquacious, impulsive inventor who needed somebody to listen to his ideas and point out the ones that couldn't possibly work. That winter, Noyce's main sounding board was his friend Gordon Moore, a thoughtful, cautious physical chemist who was another cofounder of Fairchild Semiconductor. Noyce would barge into Moore's cubicle, full of energy and excitement, and start scrawling on the blackboard: "If we built a resistor here, and the transistor over here, then maybe you could . . ."

Not suddenly, but gradually, in the first weeks of 1959, Noyce worked out a solution to the interconnections problem. On January 23, he recalled later, "all the bits and pieces came together in my head." He grabbed his notebook and wrote down an idea. It was the monolithic idea, and Noyce expressed it in words quite similar to those Jack Kilby had entered in a notebook in Dallas six months earlier: ". . . it would be desirable to make multiple devices on a single piece of silicon, in order to be able to make interconnections between devices as part of the manufacturing process, and thus reduce weight, size, etc. as well as cost per active element."

Like Kilby, Noyce felt fairly sure from the beginning that he was on to something important. "There was a tremendous motivation then to do something about the numbers barrier," he recalled later. "The [electronics] industry was in a situation—for example, in a computer with tens of thousands of components, tens of thousands of interconnections—where things were just about impossible to make. And this looked like a way to deal with that. I can remember telling Gordon one day, we might have here a solution to a real big problem."

At its core, the big problem that the monolithic idea was designed to solve was one of heightened expectations. It was hardly an unprecedented phenomenon in technological history: a major breakthrough prompts a burst of optimistic predictions about the bright new world ahead, but then problems crop up that make that rosy future unobtainable—until a new breakthrough solves the new problem.

The breakthrough that gave rise to the problem known as the tyranny of numbers was a thunderbolt that hit the world of electronics at the end of 1947. It was a seminal event of postwar science, one of those rare developments that change everything: the invention of the transistor.

Until the transistor came along, electronic devices, from the simplest AM radio to the most complex mainframe computer, were all built around vacuum tubes. Anybody old enough to have turned on a radio or television set before, say, 1964, may remember the radio tube: when you turned on the switch, you could look through the holes in the back of the set and see a bunch of orange lights begin to glow—the filaments inside the vacuum tubes. A tube gave off light because it was essentially the same thing as a light bulb; inside a vacuum sealed by a glass bulb, electric current flowed through a wire filament, heating the filament and giving off incandescent light. There has to be a vacuum inside the glass bulb or else the filament will burn up instantly from the heat. But the vacuum turned out to offer advantages that went beyond fire protection. Experimenting with light bulbs at the beginning of the twentieth century, radio pioneers found that if they ran some extra wires into that vacuum bulb, it could perform two useful electronic functions. First, it could pull a weak radio signal from an antenna and strengthen, or amplify, it enough to drive a loudspeaker, thus converting an electronic signal into sound loud enough to hear. This "amplification" function made radio, and later television, workable. Second, a properly wired light bulb, or vacuum tube, could switch—about 10,000 times in a second—from on to off. (Because of its ability to turn current on and off, the radio tube was known in England as a valve.) This capability was essential to digital computers; as we'll see in Chapter 6, computers make logical decisions and carry out mathematical computations through various combinations of on and off signals.

But vacuum tubes were big, expensive, fragile, and power hungry. They got hot, too. The lavish console radios that became the rage in the 1930s all carried warnings to owners not to leave papers near the back of the set, because the heat of all those tubes might start a fire. In a more complicated device that needed lots of tubes packed in close to each other, like a computer or a telephone switching center, all those glowing filaments gave off such enormous quantities of heat that they transformed expensive machinery into smoldering hunks of molten glass and metal—in effect, turning gold into lead. As we all know from the light bulb, vacuum tubes have an exasperating tendency to burn out at the wrong time. The University of Pennsylvania's ENIAC, the first important digital computer, never lived up to its potential because tubes kept burning out in the middle of its computations. The Army, which used ENIAC to compute artillery trajectories, finally stationed a platoon of soldiers manning grocery baskets full of tubes at strategic points around the computer; this proved little help, because the engineers could never quite tell which of the machine's 18,000 vacuum tubes had burned out at any particular time. The warmth and the soft glow of the tubes also attracted moths, which would fly through ENIAC's innards and cause short circuits. Ever since, the process of fixing computer problems has been known as debugging.

The transistor, invented two days before Christmas 1947 by William Shockley, Walter Brattain, and John Bardeen of Bell Labs, promised to eliminate all the bugs of the vacuum tube in one fell swoop. The transistor was something completely new. It was based on the physics of semiconductors—elements like silicon and germanium that have unusual electronic characteristics. The transistor performed the same two useful tasks as the vacuum tube—amplification and rapid on-off switching—by moving electronic charges along controlled paths inside a solid block of semiconductor material. There was no glass bulb, no vacuum, no warm-up time, no heat, nothing to burn out; the transistor was lighter, smaller, and faster—even the earliest models could switch from on to off about twenty times faster—than the tube it replaced.

To the electronics industry, this was a godsend. By the mid-1950s, solid state was becoming the standard state for radios, hearing aids, and most other electronic devices. The burgeoning computer industry happily embraced the transistor, as did the military, which needed small, low-power, long-lasting parts for ballistic missiles and the nascent space program. The transistor captured the popular imagination in a way no other technological achievement of the postwar era had. Contemporary scientific advances in nuclear fission, rocketry, and genetics made awesome reading in the newspapers, but were remote from daily life. The transistor, in contrast, was a breakthrough that ordinary people could use. The transistorized portable radio, introduced just in time for Christmas 1954, almost instantly became the most popular new product in retail history. It was partly synergy—pocket radios came out when a few pioneering disc jockeys were promoting a new music called rock 'n' roll—and partly sheer superiority. The first transistor radio, the Regency, was smaller, more power-efficient, far more reliable, and much cheaper ($49.95) than any radio had ever been before.

After indulging themselves for a year or two in the resentful skepticism with which academia generally greets revolutionary new concepts, physicists and electronics engineers gradually warmed to the expansive new possibilities offered by semiconductor electronics. By 1953, *The Engineering Index*, an annual compendium of scholarly monographs in technical fields, listed more than 500 papers on transistors and related semiconductor devices. A year later there were twice as many, a year after that even more. The titles reflected the intensity and the global sweep of academic research spurred by the new technology:

"Interpretation of Alpha Values in p-n Junction Transistors"

"Le Transistron dans le circuit trigger"

"Circuito Multiplicatore del coefficient di risonanza con transistor"

"Tensoranalysis in Transistor-Rueckkopplungshaltungen"

"Perekhodnaya, chastotnaya, i fazovaya kharakteristika transistora"

But the papers reflected as well a quite unscholarly enthusiasm among the academics:

"Success Story—Transistor Reliability"
"Transistors Key to Electronic Simplicity"
"Méthodes d'Optimization Appliquées à la Microminiaturization"
"Fabulous Midget"

And then, as designers learned how to make use of the fabulous midget's properties, the tyranny of numbers began to emerge. Enthusiasm gave way to frustration, even desperation. By the second half of the 1950s, when the problem was growing acute, the titles in *The Engineering Index* reflected a general sense of disappointment, even despair, in the technical community:

"Switching Losses in Transistor Circuits"
"Electronic Equipment—Weight and Volume Penalties to Flight Vehicles"
"Comment a été résolu le problème de la fabrication des transistors"
"Design Limitations of Semiconductor Components"

That last title was imprecise. The "design limitations" were not inherent in the transistors or other components; they stemmed from the basic design structure of all electric circuits.

Building a circuit is like building a sentence. There are certain standard components—nouns, verbs, adjectives in a sentence; resistors, capacitors, diodes, and transistors in a circuit—each with its own function. A *resistor* is a nozzle that restricts the flow of electricity, giving the circuit designer precise control of the current flow at any point. The volume control on a TV set is really a resistance control. Adjusting the volume adjusts a resistor; the nozzle tightens, restricting the flow of current to the speaker and thus reducing the sound level. A *capacitor* is a sponge that absorbs electrical energy and releases it, gradually or all at once, as needed. A capacitor inside a camera soaks up power from a small battery and then dumps it out in a sudden burst forceful enough to fire the flashbulb. If you have to wait until the indicator light on your camera blinks to tell you that the flash is ready to use, you're really waiting for the capacitor inside to soak up enough energy to make the thing flash. A *diode* is a dam that blocks current under some conditions and opens it to let electricity flow when the conditions change. An electric eye is a beam of light focused on a diode. A burglar who steps through the beam blocks the light to the diode, opening the dam to let current flow through to a noisy alarm. A *transistor* is a faucet. It can turn current flow on and off—and thus send digital signals pouring through the circuitry of a computer—or turn up the flow to amplify the sound coming from a radio. Just about every circuit is made of these basic parts— the nozzle, the sponge, the dam, and the faucet—which come in various speeds and sizes. By connecting these standard components in different ways, one can get circuits, or sentences, that perform different functions.

Writers of sentences are taught to keep their designs short and simple. This rule does not apply in electronics. Some of the most useful circuits are big and complicated, with hundreds of thousands of components wired together. In the era of vacuum tubes, the designers' implicit awareness of power, heat, and size restraints set a limit to the scope of any circuit design; there was no point in designing a machine that would melt to shards a few moments after it was turned on. With the transistor, though, these fundamental design limitations disappeared. Now the designers could draw up plans for exotic communications and computer circuits—circuits that could steer a rocket to the moon or control a global network of instant mail. Circuits like that might use

50,000 or 500,000 or 5 million transistors, and similar numbers of resistors, diodes, and capacitors. And why not? You could build a circuit that big now without worrying about heat or power problems. On paper, these supercircuits could outperform anything that had been designed before. All you had to do was get the parts, wire them together, and. . . . But there was the problem. That was where the numbers barrier came in. The new circuits on the drawing boards were so big and so complex it was virtually impossible to build them. Nobody could wire that many different parts together.

An electric circuit has to be a complete, unbroken path along which current can flow. That means that all the components of a circuit must be connected in a continuous loop: resistors wired to diodes, diodes to transistors, transistors to other resistors, and so on. Each component can have two, ten, even twenty interconnections with other parts of the circuit. Making the connections—wiring the parts together—was almost entirely hand labor: it was expensive, time-consuming, and inherently unreliable. A circuit with 100,000 components could easily require 1 million different soldered connections linking the components. The only machine that could make the connections was the human hand.

Even if somebody—the Pentagon, for example, where price, in the depths of the Cold War, was no object—could pay for that much hand labor, there was no way humans could put together a million of anything without turning out a few that were faulty. By the late 1950s, the electronics industry had come head-to-head with this implacable limit. The Navy's newest aircraft carriers had 350,000 electronic components, requiring millions of hand-soldered connections; the labor cost—for wiring those connections and testing each one—was greater than the total cost of the components themselves. Production of the first "second generation" (i.e., completely transistorized) computer—the Control Data CD 1604, containing 25,000 transistors, 100,000 diodes, and hundreds of thousands of resistors and capacitors—lagged hopelessly behind schedule because of the sheer difficulty of connecting the parts. And new computers on the drawing boards would be far more complex. At the end of the decade, people were already planning the computers that would someday guide a rocket to a landing on the moon. But those plans called for circuits with 10 million components. Who could produce a circuit like that? How could it fit into a rocket?

"For some time now," wrote J. A. Morton, a vice president of Bell Labs, in an article celebrating the tenth anniversary of the transistor, "electronic man has known how 'in principle' to extend greatly his visual, tactile, and mental abilities through the digital transmission and processing of all kinds of information. However, all these functions suffer from what has been called 'the tyranny of numbers.' Such systems, because of their complex digital nature, require hundreds, thousands, and sometimes tens of thousands of electron devices." "Each element must be made, tested, packed, shipped, unpacked, retested, and interconnected one-at-a-time to produce a whole system," Morton wrote in a later article. "Each element and its connections must operate reliably if the system is to function as a whole. . . . The tyranny of large systems sets up a numbers barrier to future advances if we must rely on individual discrete components for producing large systems."

In essence, the small community of engineers exploring the frontiers of electronics in the 1950s faced the same abject frustration that had confronted the small community of seamen exploring the frontiers of navigation in the 1590s. At the far western extremity of the Atlantic, hard against the shores of Central America, the explorers could look westward from the masthead and see, "with a wild surmise," a vast new ocean, a whole new world, beckoning across the isthmus. But there was no way—no way short of the impossibly expensive, time-consuming, and unreliable voyage around the tip of South America—to get to that wonderfully promising new stretch of sea. The future was within sight, tempting, tantalizing, but out of reach. Just so for Jack Kilby, Bob Noyce, and their colleagues. A vast new electronic world was right there on the blueprints, but impossible to achieve. And so physicists and electronics engineers embarked on a great voyage of discovery, searching for a route across the numbers barrier.

The search became a top-priority technological concern throughout the established world. The Royal Radar Establishment, racing to bring the honor of this important accomplishment to Great Britain, developed a promising concept as early as 1952 but failed to make it work. The French, the Germans, and the Russians competed against one another; in the United States, the Army, the Navy, and the Air Force competed just as fiercely, each service pushing its own preferred solution, each rejecting the ideas of the others. Private firms, sensing a gold mine, poured millions of dollars and man-hours into the effort. But through most of the 1950s none of these endeavors really helped. Patrick Haggerty, the president of Texas Instruments, complained that most of the proposed solutions to the tyranny of numbers "tend to exacerbate the tyranny."

The multifaceted efforts to deal with the numbers problem were grouped in the technical literature under the general title "miniaturization" (or "subminiaturization," or "microminiaturization"). It was an unfortunate term because it suggested a solution that could not work. The basic thrust of miniaturization was an effort to make electronic components extremely small, thus reducing the overall size and weight of complex electronic devices. This goal was obviously important to the military, which had to squeeze radios, radar and sonar devices, and computers into the nooks and crannies of missiles and submarines. One of the first miniaturization programs was a Navy-financed effort called Operation Tinkertoy. But there were civilian implications as well. "In civilian equipment, such as computers," the trade journal *Elec-*

tronics noted, "the number of components alone makes miniaturization essential if the computer is to be housed in a reasonable-sized building."

But turning out transistors, resistors, and the like on Tinkertoy scale did nothing to reduce the sheer number of components and connections. The Tinkertoy business tended to exacerbate the tyranny because circuits composed of tiny parts were harder, and costlier, to build. On the assembly lines, the women who soldered circuits together—it was almost entirely women's work, because male hands were considered too big, too clumsy, and too expensive for such intricate and time-consuming tasks—now had to pick up miniature components and minute lengths of wire with tweezers and join them under a magnifying glass with a soldering tool the size of a toothpick. Circuits made under those conditions were far more likely to end up with faulty connections. In many cases, even a single bad connection could be fatal to the entire circuit, just as a single burnt-out bulb can make an entire string of Christmas lights go dark. Electronic devices that relied on circuitry employing the "microminiature" parts were famously unreliable.

To enhance reliability, the designers tried redundancy. Instead of building a radio with a single set of components (a typical small radio of the late fifties might have used a half dozen transistors wired to a dozen resistors, capacitors, and diodes) the electronics companies started making radios with an extra circuit built right in—like a car built with two front axles just in case one should snap in half on the road. The redundancy approach tended to exacerbate the tyranny because the extra components and extra wire required more interconnections, and thus more labor. Worse, redundant circuitry took up more space, and that was an anathema, particularly to the people who build computers. Even without redundancy, electronic circuits were already too large. Large circuits undermined the single most important asset of modern electronic equipment: speed.

As we'll see in Chapter 9, calculators, computers, digital clocks, video games—for that matter, all digital electronic devices—are extremely dumb tools. But they are extremely fast extremely dumb tools. A computer reduces every question, every computation, every decision to the simplest possible terms (yes or no, one or zero, true or false) in the machine's internal circuitry. These two black-or-white states are represented by switches—transistors—that are either on or off. An astrophysicist mapping the universe in the observatory needs to calculate the twenty-fourth root of arctan 245.6; to do it, he types the problem into his computer. The machine has to work through a few hundred separate yes-or-no steps—that is, transistors have to switch on and off hundreds of times—just to figure out that someone has punched its keys. To determine which keys were pushed, and then to solve the problem, will take another 100,000 steps, quite possibly more. A kid playing Super Zaxxon in the arcade needs to destroy an enemy base; to do it, he pushes the "Fire" button. The machine has to work through dozens of separate yes-or-no steps just to figure out that the button was pushed. To fire the missile, and see if it hits anything, will take another 5,000 steps, quite possibly more. The machines can get away with their absurdly convoluted way of doing things only because the transistors switch, from on to off, from off to on, quickly. At a switching speed of once per second, computers would be impossible; at 1,000 times per second, merely impractical. Switching at a million times per second, computers become important. At a billion times per second—completing one step of the problem every nanosecond—they become the foundation of a revolution that has swept the world.

"After you become reconciled to the nanosecond," Robert Noyce observed, "computer operations are conceptually fairly simple." In this respect, the electronic revolution of the twentieth century is the intellectual mirror image of the biological revolution of the nineteenth. Only after they became reconciled to enormously long periods of time—enough time for a dynohippus to turn into a donkey—could Charles Darwin and his contemporaries contemplate species evolving on an evolving planet. Only after they became reconciled to enormously short periods of time—microseconds, nanoseconds, picoseconds—could the computer pioneers contemplate machines solving problems by turning switches on and off. The central concept of computer operations is that the machines operate inconceivably fast. Speed is the computer's secret weapon. If computers did not work as fast as they do, no one could justify the time and materials required to build them. At a switching speed of 1,000 times per second, it would take a whole second, maybe two, for a computer to add 2 and 2. At that rate, it would make no sense to buy the machine. The human brain, the sublimely intricate, powerful, efficient computer that everyone gets for free, can solve the problem faster than that.

The wires in an electric circuit tend to slow things down. The transistors in a computer switch on and off in response to electronic signals. A pulse of electricity moving through a wire reaches the transistor, and the transistor switches on; another pulse comes along, and the transistor switches off. No matter how quickly the transistor itself can switch, it cannot do so until the pulse arrives telling it what to do. The more wiring there is in a circuit, the farther these messenger pulses have to travel. In the 1950s, the limiting factor in computer speed was the travel time for those electronic signals moving through the circuit. In the biggest computers, with literally miles of wiring, it took so long for pulses to travel from one side of the circuit to the other that computation rates were seriously impaired.

At first blush, it might appear that there were two potential solutions to this problem: either speed up the signals, so they move through a large circuit faster, or shrink the circuits. Someday, if relativity theory is displaced and the laws of twentieth-century physics are stood on their heads, the first solution may be at hand. At present, however, it is against the law. Electronic signals move through a circuit at the universal speed limit—the speed of light. If modern physics is correct, nothing will ever move faster. That leaves the second solution. To increase computing speed, it was necessary to reduce the distance the messenger pulses had to travel—that is, to make the circuits smaller. But smaller circuits meant decreased capacity. The result was a paradox. In the argot of the engineers, a computer's "power" is a measure of both its capacity to handle big problems and its speed in solving them. It was possible—at least in applications where cost and size were not serious problems—to increase problem-solving capacity by wiring in more transistors. But more transistors and more wire meant a larger circuit, which reduced computing speed. Thus the effort to build in increased computing power led to decreased computing power. It was the technological equivalent of Catch-22: the tyranny of numbers.

"It was a situation where, quite clearly, size dictated performance," Bob Noyce recalled. "Not just performance, in the sense of limiting computing speed, but the size and complexity of electronic circuits dictated cost, reliability, utility."

"The things that you could see that you wanted to do were going to take so many transistors, so many parts, that it would just be prohibitive, from a cost standpoint, from a size standpoint, any way you wanted to look at it," Jack Kilby remembered.

Noyce: "A large segment of the technical community was on the lookout for a solution."

Kilby: "There was just an awful lot going on. . . . It was pretty well accepted that this was the problem that had to be solved."

Noyce: "It was clear that a ready market awaited the successful inventor."

.   .   .

As it happened, the successful inventors were a pair of engineers working at competing manufacturing companies. That is to say, this global technological problem of earth-shaking importance was not turned over to scientists. It was not solved by academic researchers in a university laboratory, but rather by engineers at industrial lab benches just down the hall from the production line.

Scientists and engineers tend to divide their work into two large categories, sometimes described as basic research and directed research. Some of the most crucial inventions and discoveries of the modern world have come about through basic research — that is, work that was not directed toward any particular use. Albert Einstein's picture of the universe, Alexander Fleming's discovery of penicillin, Niels Bohr's blueprint of the atomic nucleus, the Watson-Crick "double helix" model of DNA — all these have had enormous practical implications, but they all came out of basic research. There are just as many basic tools of modern life — the electric light, the telephone, vitamin pills, the Internet — that resulted from a clearly focused effort to solve a particular problem. In a sense, this distinction between basic and directed research encompasses the difference between science and engineering. Scientists, on the whole, are driven by the thirst for knowledge; their motivation, as the Nobel laureate Richard Feynman put it, is "the joy of finding things out." Engineers, in contrast, are solution-driven. Their joy is making things work.

The monolithic idea was an engineering solution. It worked around the tyranny of numbers by reducing the numbers to one: a complete circuit would consist of just one part — a single ("monolithic") block of semiconductor material containing all the components and all the interconnections of the most complex circuit designs. The tangible product of that idea, known to engineers as the monolithic integrated circuit and to the world at large as the semiconductor chip, has changed the world as fundamentally as did the telephone, the light bulb, and the horseless carriage. The integrated circuit is the heart of clocks, computers, cameras, and calculators, of pacemakers and Palm Pilots, of deep-space probes and deep-sea sensors, of toasters, typewriters, cell phones, and Internet servers. The National Academy of Sciences declared the integrated circuit the progenitor of the "Second Industrial Revolution." The first Industrial Revolution enhanced man's physical prowess and freed people from the drudgery of backbreaking manual labor; the revolution spawned by the chip enhances our intellectual prowess and frees people from the drudgery of mind-numbing computational labor. A British physicist, Sir Ieuan Madlock, Her Majesty's Chief Science Advisor, called the integrated circuit "the most remarkable technology ever to hit mankind." A California businessman, Jerry Sanders, founder of Advanced Micro Devices, Inc., offered a more pointed assessment: "Integrated circuits are the crude oil of the eighties."

All this came about because two young Americans came up with a new idea—or, more precisely, a not-so-new idea. In fact, the principle underlying the semiconductor revolution was one of the oldest ideas in electronics.

2

THE WILL TO THINK

In addition to the laws, rules, constants, principles, axioms, theories, and hypotheses they have devised to explain the mysteries of the natural world, scientists and engineers have developed a series of unwritten rules that purport to explain the mysteries of their business. Among the latter is a humorous, or perhaps quasi-humorous, principle sometimes referred to as "the law of the most famous." Briefly put, this natural law holds that whenever a group of investigators makes an important discovery, the most famous member of the group will get all the credit. And that seems to explain why the important principle of thermionic emission came to be known as the Edison Effect.

Thermionic emission was observed for the first time in March 1883 in Thomas A. Edison's Menlo Park laboratory, when the inventor and his associates noticed something strange going on inside one of his first light bulbs. In addition to the electric current flowing through the carbon filament, there seemed to be another, separate current flowing through the vacuum inside the glass bulb—something quite impossible, under contemporary explanations of electricity. Nobody at Menlo Park understood what was happening (the current was eventually found to be a flow of electrons boiling off the white-hot filament). But Edison, not one to miss a chance, wrote up the discovery and filed a patent for it. Then he set it aside. It was partly a matter of time and resources. Bedeviled by disputes with his creditors, legal battles over his patents, and frustrating efforts to improve his phonograph, microphone, and incandescent lamp, Edison was already working twenty hours a day, often more. But the main reason Edison abandoned the Edison Effect was that he saw no future in it. So what if current could flow through a vacuum? What good would that do?

Looking back 120 years later, when solid-state, or semiconductor, devices have proven superior to vacuum tubes in just about all electronic equipment, we can see that Edison had a point. The science and industry of electronics was based on the thermionic effect inside a vacuum tube for more than fifty years; today, hindsight suggests that the half century of work on the vacuum tube was basically a digression. By the time Edison discovered his eponymous effect in 1883, electrical pioneers such as Edmond Becquerel, Ferdinand Braun, and Michael Faraday had already found that certain substances—known today as semiconductors—had a variety of useful electronic characteristics. Had the early work on these materials been continued, it is not too great a flight of fancy to suggest that the modern semiconductor revolution might have come a half century or more earlier than it did. But after the discovery of the Edison Effect, electronics research took a new direction—with dramatic results. The work at Menlo Park led, fourteen years later, to the experiment known as "the zero hour of modern physics"—the discovery of the electron—and from there, along a more or less straight line, to wireless telegraphy, radio, television, and the first generation of digital computers. It was all a digression, but a glorious one.

In 1883, however, that point was considerably less than obvious, even to an exceptional visionary like Edison. At first, the Edison Effect held interest only for scientists—and scientific interest was not a commodity of great importance to Thomas A. Edison.

"Well, I'm not a scientist," the Wizard of Menlo Park said. "I measure everything I do by the size of the silver dollar. If it don't come up to that standard then I know it's no good."

That line was classic Edison. All his life he portrayed himself as the supreme pragmatist, relying on hard work and common sense to build a better life for his fellow man—and get rich in the process. It was an archetypal American picture, and essentially an accurate one, for Edison's life has the elements of the classic American story. The seventh child of an infrequently successful businessman, he grew up in medium-size towns in Ohio and Michigan. He had about four years of school, counting the time his mother tutored him at home, and set out at the age of twelve to make his fortune. He sold snacks on the Detroit–Port Huron train. He started a newspaper called *Paul Pry*. He fell into and out of numerous jobs as a telegraph operator, a profession he enjoyed so much that he called his first two children Dot and Dash. Tinkering continually with his employers' telegraphic equipment, he began to design useful improvements, and get paid for them. Gradually, he discovered that he could make a living as an inventor. By his thirty-fifth birthday, Edison was a millionaire, a leader of industry, and probably the best-known man on earth. When he announced early in 1878 that he might try to perfect an electric light, illuminating gas stocks plummeted on Wall Street. When the *New York Daily Graphic* reported that Edison had invented a machine that spun food and wine from mud and water, many newspapers failed to notice the April 1 dateline and ran the story straight—just one more miracle from Menlo Park. When Edison died, at eighty-four, in 1931, someone proposed that all the lights in the world be turned out for two minutes as a memorial. The idea was dropped on the ground that it would be impossible for the world to function that long without the electric light.

Despite fame and fortune, Edison remained an uncouth hayseed who flaunted his disdain for cleanliness, fashion, order, religion, and science. A journalist touring the famous Menlo Park laboratory in 1878 described the proprietor this way: "The hair, beginning to be touched with gray, falls over the forehead in a mop. The hands are stained with acid, his clothing is 'ready-made.' He has the air of a mechanic, or with his particular pallor, of a night-printer." Edison worked in 40-, 80-, or 100-hour bursts, catching a few intervals of sleep under a bench in the lab. He adopted the motto *Perseverantia omnia vincit*, a phrase that he subsequently translated to "Genius is 1 percent inspiration and 99 percent perspiration." He had absolute confidence in this formula; he was certain he could solve any problem if he just tried enough solutions. After an assistant noted wearily that the Menlo Park team had worked through 8,000 different formulas in a futile effort to build a storage battery, the inventor replied that "at least we know 8,000 things that don't work." Struggling to find an efficient filament for his incandescent light, Edison decided to try everything on earth until something worked. He made a filament from dried grass, but that went haywire. He tried aluminum, platinum, tungsten, tree bark, cat's gut, horse's hoof, man's beard, and some 6,000 vegetable growths before finding a solution in a carbonized length of cotton thread.

Publicly, at least, the great empiricist had no truck with scientists, mathematicians, or any college graduate—"filled up with Latin, philosophy, and all that ninny stuff." With great fanfare, he invented an "ignoramer" to prove that intellectuals had no common sense. When he hired a "science advisor" of his own— Francis Upton, holder of a Ph.D. from Princeton and student of the great German physicist Hermann Helmholtz—Edison promptly named him "Culture," a term of derision among the gang at Menlo Park. Soon enough the newspapers were reporting how Edison had asked his man Culture to determine the volume, in cubic centimeters, of an empty glass bulb. Upton set to the task with a complex series of differential equations that would lead, after a few hours, to a close approximation of the correct figure. With a smirk, the pragmatic inventor grabbed the bulb, filled it with water, and poured the water into a measuring cup— determining the precise volume in less than a minute.

Culture eventually became acculturated to life in the lab, and Edison eventually came to realize that Upton's mathematical and physical skills were an important asset. Upton designed and built the dynamo—in modern terms, the power plant—that provided safe, efficient power for the Edison lighting system and the Edison electric train. Upton's skills as an observer and theoretician were so central to the discovery of thermionic emission that the phenomenon might more fairly be called the Upton Effect.

Edison's electric lamp depended, of course, on a clear glass bulb. The bulb enclosed a vacuum. In that vacuum, electricity running through the filament caused the filament to glow with a white heat (the vacuum was necessary because without air, the filament would not burn away). The glow showed through the glass bulb and gave light. When the electric light was still in its birth throes, Edison and Upton noticed that, over time, the clear glass bulb tended to grow black. This was a problem, and the inventor set out in his standard way to eliminate the problem. He launched into an exhaustive series of trial-and-error experiments to find out what was wrong. Upton, meanwhile, made careful observations of the phenomenon. In March 1883 he suggested putting a small metal plate inside the bulb. This didn't help (the problem of darkening bulbs was eventually solved by using purer materials for the filament), but it led to a fascinating discovery. Being scientists, being naturally curious people, they spent some time looking over their failed idea—that is, the light bulb with a metal plate in it. They found, to their surprise, that electric current was flowing in the metal plate. Where in the world could this current come from? Current was flowing through the filament, of course, but there was no connection whatsoever between the filament and the plate. Being naturally curious people, the researchers tried some tests. They increased the current through the filament—and found that the current in the metal plate increased proportionally. Evidently, electric current was flowing across the vacuum from filament to plate. This was quite astonishing—at least to Upton, who trusted scientists—because the experts had established incontrovertibly that electric current could not traverse a vacuum. Now Edison had proven them wrong.

Over the next few years, the mysterious current-in-a-vacuum became a prize exhibit at electrical exhibitions on both sides of the Atlantic. Given its famous discoverer, the impossible current was quickly dubbed the Edison Effect. John A. Fleming, a scientist on the staff of Edison's British subsidiary, ordered some Edison Effect lamps and tried a number of experiments. Edison had always used a direct current of electricity in his work—a current in which the electricity flowed in the same direction all the time. During the winter of 1884–85, Fleming tried something different. He hooked up the filament to a generator that produced alternating current—a current that constantly changes direction, back and forth, back and forth, as often as 120 times per second. He was mystified to see that, even with an alternating current racing back and forth through the filament, the current flowing to the metal plate was still direct current, never changing direction. The lamp had converted alternating current to direct current.

No one could explain this result—least of all Edison, of course, who stood aside and smiled as the scientists did their stuff. Research on the Edison Effect was merely aesthetics, Edison wrote to a friend, and "I have never had time to go into the aesthetic part of my work. . . . But it has, I am told, a very important bearing on some laws now being formulated by the Bulged-headed fraternity of the Savanic world."

It did indeed. Savants in the United States and Europe undertook extensive experimentation on the flow of electricity through a vacuum. The basic apparatus for this work was an elongated glass tube with a piece of carbon or metal at one end that was heated—just like the filament in Edison's light bulb—until the thermal energy emitted an electric current (hence "thermionic emission") through the vacuum. The piece of metal that emitted the current was called a cathode, so the glass tube was known as a cathode ray tube. The current would beam down the tube to the far end; at the spot where the beam hit, the glass would phosphoresce, or glow. At the time, this exotic piece of scientific apparatus was found only in the finest laboratories in the United States and Europe. Today it is found in living rooms, basements, and bars everywhere, in the form of the television picture tube.

How a cathode ray tube works.

The savants hoped that the tube would provide a clear enough picture of cathode rays to permit an explanation of the electric force. As the nineteenth century neared an end, there was a curious gap between the engineers and the scientists. Thanks to the engineers and pragmatic inventors like Edison, electricity powered much of the world and made the nights shine as day. But the scientists still didn't know precisely what this mighty force was. The mystery of electricity had prompted a number of contradictory hypotheses. Early researchers had postulated that electricity was a fluid (which is why we still talk today of "current" and "flow"). Then, in the 1880s, this notion gave way to a pair of competing theories. One view held that electricity was a wave phenomenon, like sound and light; the other school of thought considered the electric beam in the cathode ray tube to be a stream of particles, like grains of sand. Wave or particle? The greatest minds in physics pondered, debated, speculated over the question. The answer finally came from one of the most fascinating and formidable intellects in the history of physics—Professor Sir Joseph John Thomson.

J. J. Thomson was born in 1856, the son of a bookseller in Manchester. The family's plan had been that J.J. would be apprenticed to a local engineer and take up that profession. But his father's death, when J.J. was sixteen, left the Thomsons unable to pay the fees engineers charged for training apprentices. The boy won a scholarship at a local college and quickly came under the spell of mathematics and physics. The timing was perfect—Thomson's professional career spanned the most fertile era in physics since Isaac Newton's day—but nobody knew that during his college days. Quite the contrary, in fact: in the 1870s, there was a general sense that the interesting part of physics was over, and all that remained was refining the measurements. After all, everyone knew that all matter was made of indivisible particles called atoms, and that differences among elements were due to differences in their atoms. The great theoretician James Clerk Maxwell, on the august occasion of his nomination as Cavendish professor at Cambridge—in essence, the physicist laureate of England—noted the consensus "that in a few years all the great physical constraints will have been approximately estimated, and that the only occupation which will then be left to men of science will be to carry on those measurements to another place of decimals." Already scientists had measured the mass of the smallest object in the universe—the hydrogen atom, weighing about .00000000000000000000000017 gram (which is to say, about .00000000000000000000000006 ounce).

Thomson went to Cambridge in 1876, studied under the great Maxwell, and quickly distinguished himself. He won the Trinity Prize for the interesting discovery that a drop of water would not evaporate if given an electric charge. In an 1881 paper he suggested that there was a connection between the mass and the energy of a moving sphere—an idea that was crystallized by another physicist twenty-four years later in the equation $E = MC^2$. As the preeminent Cambridge physicist, Thomson succeeded in 1884 to the chair—both the professorship and the piece of furniture— that Maxwell had held. He was twenty-eight years old. "Things have come to a pretty pass," an older colleague groused, "when mere boys are appointed professors." The mere boy took over Maxwell's old office in the Cavendish Laboratory, a cream yellow neo-Gothic pile with the Cavendish family's motto emblazoned over the door: *Cavendo tutus,* or "Always on the Lookout."

"J.J. spent a good part of most days sitting in the armchair that had belonged to Maxwell, doing mathematics," a student recalled. He sat there thinking, working out problems, because he had absolute confidence in the power of thought. He was certain that the human mind, aided by mathematics, could comprehend all physical phenomena. But the professor's wide-ranging curiosity reached far beyond math and physics. His workday ran, at the most, from ten in the morning to six at night (with a break for afternoon tea), and he regularly found time to cheer for Cambridge at crew and rugby matches. He read A. E. Housman's poetry in manuscript and made it a point never to miss opening night of a new Gilbert and Sullivan operetta. He was fascinated by gardening, golf, and American politics. He was British to the core. In his memoirs he notes with great pride that twenty-seven of his students (including his son) were elected to the Royal Academy; as an aside, he mentions that seven of them (including his son) also picked up Nobel Prizes. J.J.'s own Nobel Prize, in 1906, seems to have satisfied him less than the knighthood he received two years later. When he died, at eighty-four, in 1940, he was buried in Westminster Abbey near the grave of Isaac Newton.

Physicists talk about Thomson's cathode ray experiment of 1897 the way architects talk about the Guggenheim Museum in Bilbao. The approach was elegant, the conclusion bold and stunning, and the result left an indelible mark on everything that came after. Thomson was determined to find out everything he could about that electric beam shooting through the cathode ray tube. Where did it come from? What was it made of? Was that beam of pure energy a wave, or a stream of particles? Thomson decided he could ascertain the nature of the electric beam by subjecting it to different forces and measuring the results as precisely as possible. His experiments were beautiful to watch. By placing magnets around the glass tube, Thomson could make the cathode ray spin in a perfect spiral. Since the vacuum in the tube was not complete, the cathode ray would occasionally strike a stray gas atom and give it an electric charge. The charged atom, or ion, would glow with a colorful splendor until it was neutralized, or recombined, by another atom. So lovely were the results that Thomson and his colleagues wrote a song about it all, to the tune of "Clementine":

> *In the dusty lab'ratory,*
> *'Mid the coils and wax and twine,*
> *There the atoms in their glory*
> *Ionize and recombine.*

> *Chorus: Oh my darlings! Oh my darlings!*
>   *Oh my darling ions mine!*
>   *You are lost and gone forever*
>   *When just once you recombine.*

In the weird magnetic circuit
See how lovingly they twine,
As each ion describes a spiral
Round its own magnetic line.

Chorus

In a tube quite electrodeless,
They discharge around a line,
And the glow they leave behind them
Is quite corking for a time.

Chorus

At first, Thomson's research produced nothing but confusion; his findings did not seem to fit any single theory. The fact that the electric beam could be bent by a magnet or an electric field strongly suggested that it consisted of particles, for no wave was susceptible to magnetic attraction. But there was also compelling evidence against the particle hypothesis. There was no indication, even with the most exacting measurements, that the beam was deflected by gravity, as any stream of particles should be. Further, the beam could pass through a thin sheet of metal foil without leaving a hole. It was known that certain energy waves did that— just as light waves pass through a window without leaving a mark on the glass—but it seemed impossible for a solid particle to do so. To resolve these contradictory results, Thomson fell back on mathematics. Through a series of ingenious calculations, he determined the velocity of the moving beam—it traveled about 20,000 miles per second, infinitely faster than any object had ever been found to move before. Using that result, he calculated what the mass of the cathode ray particles would be, if indeed they were particles. This produced a result that was simply impossible. The mass of each particle would be less than one one-thousandth the size of a hydrogen atom. As a physical matter, this could not be right, for the hydrogen atom was well known to be the smallest of all material objects, and quite indivisible. As a mathematical matter, though, it could not be wrong. At this point, Thomson ended his experiments. There was nothing left to do but sit down in Maxwell's armchair and think the matter through.

The conclusion Thomson drew from this research is so elementary today that it is almost impossible to appreciate the enormous intuitive leap, the sheer revolutionary daring, that was required to state it in 1897. In one fell swoop, he split the unsplittable atom and explained the inexplicable force of electricity. He presented his wholly new picture of the physical world in an informal speech to the Royal Institution on a Friday evening in April.

"I have lately made some experiments which are interesting," he began. Contrary to accepted scientific wisdom, he went on, the atom was not indivisible. He had found within the atom a new kind of particle—a particle at least a thousand times smaller than any atom. These subatomic particles—Thomson called them "corpuscles," but eventually the term "electrons" was adopted instead—are universal constituents of all matter, found in every atom. The electrons are magnetic and carry a negative charge, which explains why a magnet or an electric field made them swerve from their straight path. Their mass is so minute that the force of gravity upon them was undetectable. They are so much smaller than any atom that they could slip through the open spaces in an atom of metal—and thus shoot through a sheet of metal foil without a trace. The cathode ray—for that matter, any electric current—consists of a stream of these charged particles.

In subsequent lectures and papers, Thomson and his colleagues refined their picture of the electron and postulated other subatomic particles. Since the electrons carried a negative charge, there must be another, positively charged particle in the atom; equal numbers of electrons and protons (as the positive particles came to be called) would make the ordinary atom electrically neutral. Because of its small mass, the electron can move about. Indeed, if a piece of metal is given an electric charge or heated to incandescence (as in the Edison Effect), electrons will stream away in huge numbers. This stream of moving electrons, Thomson concluded, is an electric current.

As with any shocking discovery, J. J. Thomson's conclusion that the atom consisted of many smaller parts drew resistance at first. Fairly soon, though, the world of physics grew to accept the existence of the electron and the proton because the theory worked. It explained all sorts of things about atomic structure and about electricity that had not been understood before. It would eventually make possible the whole new world of semiconductor physics. It was a seminal idea.

The cream yellow laboratory where this idea was born still stands on the Cambridge campus, although nowadays it houses political scientists and economists. The university has built a fancy new Cavendish Laboratory for physicists and engineers a fair distance away. But at least the trustees of Cambridge have a sense of history. They marked the 100th anniversary of Thomson's great breakthrough by mounting a celebratory plaque on the wall, outside his old lab:

Here in 1897 at the old Cavendish Laboratory
J. J. THOMSON
discovered the electron
subsequently recognized as the
first fundamental particle of physics
and the basis of chemical bonding, electronics, and computing.

Although Thomson's leap of insight developed from experiments on electric current in a vacuum, his explanation of electric

current—a flow of moving charges—is equally valid for current in
a block of semiconductor material. If scientists had gone back to
the semiconductor work that had largely been laid aside a generation earlier, the next important developments might have come
in that field. As it happened, though, Thomson's discovery was
first applied in vacuum tubes—and the tube became the central
component of electronic devices for the next half century. This
came about because of the work of John Ambrose Fleming and
Lee De Forest.

J. A. Fleming was a contemporary of Thomson's. Like Thomson, he studied physics at college only because his family could
not pay for an engineering apprenticeship. Like Thomson, Fleming was a distinguished professor (at University College, London)
and a member of the leading scientific societies. Unlike Thomson, Fleming was interested in making money from his work; one
result was that, like Edison, he became ensnarled in endless
patent litigation. In pursuit of an income, Fleming worked the
lecture circuit, traveling all over the British Isles to give scientific
demonstrations. Hard-working, highly disciplined, extremely demanding of himself and those around him, Fleming was determined that everything about his lectures should be perfect—he
rehearsed with a stopwatch so that every word and gesture would
come at the right second—and thus failed to see the humor in a
prank perpetrated during a widely advertised talk he gave in 1903.
To demonstrate the wonders of wireless telegraphy, Fleming had
arranged to receive a long-distance message in mid-lecture from
the great Guglielmo Marconi himself. At the appropriate time,
the telegraph key began to rattle, but instead of the weighty words
Fleming had chosen to mark the historic occasion, the message
was an off-color limerick ("There was a young fellow of Italy/
Who diddled the public quite prettily . . ."). It turned out that a
playful student in the audience, a precursor of the modern
hacker, had brought a transmitter of his own for purposes of mis-

chief and broken into the network. Fleming, thoroughly una-
mused, wrote a thundering letter to *The Times* to denounce such
"scientific hooliganism."

Fleming's interest in money also led him to pioneer a practice
common among modern faculty members—consulting to industrial concerns. In 1882 he was appointed "electrician" (the modern term would be "science advisor") to the Edison Electric Light
Company, Ltd. He held that position only a few years, but they
happened to be the years when Edison and Upton were working
on the Edison Effect. Fleming, as we have seen, made the interesting discovery in 1884 that the Edison Effect current—the current from the filament to the metal plate—never changed
direction, even when alternating current was sent through the filament. Years later, after Thomson had established that the current is a flow of electrons, Fleming was able to explain why.
Electrons boiling off the hot filament flowed to the metal plate.
But the plate was not hot enough to emit electrons, so no current
flowed back from plate to filament. Thus the Edison Effect always
produced direct current.

At the turn of the century Fleming landed another consulting
position, this time with Marconi's Wireless Telegraph Company,
Ltd. Radio in this primordial era was still as much a toy as a tool,
and there were several problems facing the Marconi firm. One
was the inability to tune the radios to a specific frequency, an improvement that could have prevented pranksters from sending
limericks in place of important messages. A more significant obstacle to serious radio transmission was the absence of a reliable
rectifier. A radio transmitter beams out signals that travel through
the sky in the form of alternating current. But the receiving instruments that turn those signals into information—a telegraph
key, for example, or a radio's speaker—operate on direct current.
The crucial need, then—the missing link—was a device that
could take the alternating current sent by the transmitter and con-

vert, or rectify, it to a direct current that echoed the pulsations in the original signal.

The materials known today as semiconductors have this rectifying quality, and the first radios employed a semiconductor crystal to rectify the signal. Such radios came to be known as crystal sets. Since nobody knew much about semiconductor technology, crystal sets seemed to work only when they chose to, and were too capricious for business use. If radio was to have any serious impact on the world, someone would have to find a dependable way to convert the alternating radio signal into direct current. This was the task Marconi assigned to J. A. Fleming.

Fleming initially tried to get reliable rectifying action out of the standard crystal. In October 1904, however, he realized that this approach would not be fruitful and began thinking hard about other devices that could convert an alternating current to direct current. What mechanism always produced direct current? Suddenly, he recalled his experiments of twenty years before. Fleming himself described the moment of discovery:

> I was pondering on the difficulties of the problem when my thoughts recurred to my experiments in connection with the Edison Effect.
>
> "Why not try the lamps?" I thought.
>
> I went to a cabinet and brought out some lamps I had used in my previous investigations. . . . I started the oscillations in the primary circuit. To my delight I saw the needle of the galvanometer indicate a steady direct current. . . . We had in this particular kind of electric lamp a solution to the problem of rectifying high frequency wireless currents. The missing link in wireless was found—and it was an electric lamp.

In addition to its usefulness for radio, Fleming noted another important characteristic: the current flowing from filament to plate could switch off and on far more rapidly than any mechanical switch. "So nimble are these little electrons," Fleming wrote, "that however rapidly we change the electrification . . . the plate current is correspondingly altered, even at the rate of a million times per second." Because of this capacity to turn current on and off like a faucet, Fleming called the modified light bulb a "valve." In the technical literature, Fleming's rectifying lamp is called a diode, because it has two electrodes—the filament and the plate.

The Fleming diode made possible the production of dependable radio receivers. But there was still another problem to be resolved before radio became a practical instrument for sending information over any appreciable distance. Radio beams attenuate as they travel; the farther a signal has to go, the weaker it gets. After a signal had traveled 30 miles or so it was too weak to drive any kind of microphone; a slightly longer distance so diminished the current that it could barely move a telegraph key. What was needed was a device in the receiver that could strengthen, or amplify, the incoming signal without distorting its pattern of pulsation and modulation. The need was met, two years after Fleming's invention, in the cluttered New York office of Lee De Forest.

De Forest was the son of a Congregationalist minister who moved, shortly after the Civil War, from the Midwest to Talladega, Alabama, to run the Negro college there. Stuck with this double whammy—a Yankee who lived with the Negroes—young Lee had few friends among his fellow whites in Talladega and spent his childhood reading science. He did indeed become a scientist; he took a Ph.D. at Yale after writing what was probably the first American dissertation on radio waves. Still, there was considerably more of Edison in him than of Thomson. A tireless self-promoter, De Forest worked feverishly all his life to attain the wealth and fame he felt he deserved. It was often an uphill battle. De Forest spent huge sums, with indifferent results, in legal battles over patents. He spent two years at the height of his career, 1912–13, fighting a federal mail fraud indictment resulting from his prediction, in a letter to potential investors, that the human voice would someday be broadcast across the Atlantic. No one could possibly believe such "absurd and deliberately misleading statements," the prosecutor declared. The jury did, and De Forest was acquitted.

Not a shrinking violet, De Forest yielded to none in his esteem for his own scientific accomplishments. Asked to discuss his radio amplifier before the Franklin Institute, the inventor assured the assembled engineers that "a more revolutionary step was never taken in the history of engineering." De Forest titled his autobiography *Father of Radio*, an immodesty that prompted Isaac Asimov to note that "few inventions have had so many fathers." Just before he underwent delicate cancer surgery at the age of eighty-five, De Forest overheard the doctors saying the tumor would be removed by electrodesiccation. "Commonly known as the hot wire," De Forest croaked from the operating table. "I invented it in 1907."

De Forest's most important invention, the radio amplifier, was based on a fundamental principle of electricity: unlike charges attract, and like charges repel. An object carrying a negative charge is attracted toward a positive charge just as a paper clip is pulled toward a magnet. In a rainstorm, when clouds and earth develop opposite charges, the attraction is strong enough to pull a lightning bolt of electrons across the gap. This principle also explains the paparazzi's favorite phenomenon, static cling. As the starlet scoots over to get out of the limousine, the hem of her dress rubs some electrons off her nylon stockings. With these excess electrons, the dress acquires a negative charge; the stockings become positive. Negative hem clings to positive stocking—at about midthigh, if the photographers are lucky—and a thousand flashbulbs pop as she leaves the car.

Tinkering with some early rectifying tubes in 1906, De Forest put a wire screen between the filament and the metal plate. Normally, this did not affect the Edison Effect current because electrons flowed right through the open screen to the plate. But when De Forest put a negative charge on the wire screen, like charges repelled: the negative screen repelled the negatively charged electrons, and current flowing to the plate was sharply reduced. When he sent a positive charge to the screen, unlike charges attracted: the screen attracted electrons, and the current to the plate was increased.

Experiments showed that a small change in the charge on the metal screen caused a big change in the current flowing to the metal plate. More important, the variations in the Edison Effect current exactly mimicked variations in the current sent to the wire screen. Here, then, was a precise amplifying mechanism. If the weak current from a distant radio signal was sent to the wire screen, it shaped a much stronger current that precisely matched the fluctuations of the radio beam. This stronger current could drive a telegraph key or loudspeaker. De Forest called the metal screen a "grid" and filed a patent entitled "Device for Amplifying Feeble Electrical Currents."

Since the De Forest tube had three electrodes—the filament, the metal plate, and the grid—it was technically known as a triode. The triode amplifier made radio a practical everyday reality. With further advances—radio companies eventually developed tetrode and pentode tubes—performance was greatly enhanced. By 1930, radio had swept the world (De Forest's absurd prediction of transatlantic broadcasting came true in 1915). By the mid-forties, engineers had learned how to use radio signals to draw a picture on a glass tube; the result was a device known at first as an iconoscope but today as television. Computer pioneers took advantage of the tube's rapid switching in building the first generation of digital computers.

With the capacity to perform three essential functions—rectification, amplification, and rapid switching—the vacuum tube, at heart just a souped-up light bulb, was the hub of a new electronic world. If the development of electronics were viewed as a battle of competing technologies, vacuum tubes had overcome semiconductor devices, like the crystal set, and left them far behind. But this battle was not yet over.

The renaissance of the semiconductor began in the late 1930s, spurred by growing fear on both sides of the English Channel that war was imminent. Recognizing that the coming conflict would depend largely on airpower, scientists in England and Germany raced to develop an early warning system, using radio techniques, to spot approaching enemy planes. Fortunately for the Allies, the British perfected the concept first. Since their system could not only find planes but also gauge their distance, or range, from England, the British called the invention Radio Detection and Ranging. This name was quickly shortened to the acronym "radar."

A radar station shot a radio beam into the air. As long as it didn't run into anything, the beam kept moving in a straight line at the speed of light (some radar beams sent off during the Battle of Britain are presumably still moving out through space today, about 61 light-years from earth). But if the signal hit a piece of metal—say, a Luftwaffe bomber—in midair, the beam would bounce back to the radar station like a tennis ball bouncing off a wall. By marking where the returning beam came from, and measuring how long its round trip had taken, the British defenders could tell their fighters where to intercept the enemy.

At first the British had hoped to use standard radio equipment to transmit and receive the radar beams. This didn't work, because vacuum tube rectifiers could not handle the high-frequency signals required for radar. Desperate for some other rectifying apparatus, the engineers went backward in history and resurrected the crystal set. Crystal rectifiers had been around for decades, but they had never been significant because their performance was always iffy. By the late 1930s, however, much more was known about the crystals in these crystal sets — the elements known as semiconductors. The radar engineers, working from this new base of knowledge, were able to build reliable crystal receivers.

The deeper understanding of semiconductors had come about in a random, almost haphazard, manner. Unlike the development of vacuum tube technology, in which each new researcher and each new discovery seemed to lead neatly ahead to the next, the scientific world's knowledge of semiconductors grew out of a disjointed series of experiments and hypotheses. A monograph published in Berlin, an interesting experiment in Cambridge, a suggestion from Paris, a countertheory from Copenhagen — all this work gradually came into focus in the 1930s. Eventually, two lines of scientific work, one experimental and one theoretical, merged into a single theory of semiconductor physics.

The experimental contribution sprang from commercial roots. As electrical power became a commercially important commodity, the firms that sold and used electricity had to know the most efficient way to transmit it. If a company had to build a power line, should the line be made of copper or cotton? To answer that, experiments were run to measure the conductivity of countless different materials. Conductivity — that is, how easy it is for an electric current to flow through a given material — is a physicist's concept; it is the opposite of the electrical engineer's concept of resistance. The electrician's unit of resistance is called the ohm. Some physicist with a wry humor accordingly decided that the unit of conductivity should be called the "mho."

In experiments to determine mho ratings for hundreds of materials, certain patterns emerged. In some substances, particularly metals like silver, gold, and copper, current could flow easily.

These materials were labeled conductors. Other substances—quartz, glass, rubber, and wood are prominent examples—blocked current flow. They were called insulators. Between these two extremes was a class of materials that conduct better than insulators but not as well as conductors. These semi-good conductors—elements like selenium, germanium, and silicon—were given the generic label "semiconductors." Still lacking, though, was an explanation of why some materials made better conductors than others. Why did electrons travel so readily through copper but so reluctantly through glass? And what was it about silicon that made it fall in between? The answer to that question was provided by the theorists of quantum mechanics, particularly by a quiet, deferential Dane, Niels Henrik David Bohr, who worked out the basic architecture of the atom.

Niels Bohr was one of the great people of the twentieth century—not only one of the most powerful intellects but also one of the most generous and humane of men: "the incarnation of altruism," his friend C. P. Snow wrote. Born in Copenhagen in 1885, the son and grandson of distinguished academicians, he grew up in highly intellectual surroundings. The family read widely in four languages and was steeped in music and the arts. Niels was also a soccer ace, although he was not picked for the Danish Olympic team (his brother snared a spot in 1908). Denmark did come beckoning a few years later, however, when Bohr had emerged in the top ranks of physicists and was working in England. To lure Bohr home, the Danes established an Institute of Theoretical Physics in Copenhagen; as it almost always does in Denmark, the money came from the Carlsberg Brewery. Under Bohr's direction the institute became, for a while, the world capital of quantum physics.

"He was not, as Einstein was, impersonally kind to the human race," C. P. Snow wrote after Bohr's death in 1962. "He was simply and genuinely kind. It sounds insipid, but in addition to wisdom he had much sweetness." His selfless nature shone through radiantly during World War II. He donated a priceless possession—his gold Nobel Prize medal—to be melted down so that the proceeds could go to Finnish war relief. At considerable professional and personal danger, he spoke out forthrightly against the Nazis and worked secretly to help Jewish scientists escape the Third Reich. When he learned that his most brilliant former student, Werner Heisenberg, had been put in charge of Germany's research program to develop an atomic bomb, Bohr summoned the younger physicist from Germany and told him that it would be wrong, morally wrong, for any scientist to give Hitler this weapon. (Whether or not that lecture from Bohr was the reason, Heisenberg made minimal progress at best on the Nazi nuclear program.) Finally forced to flee his homeland in a small fishing boat with his son Aage (a chip off the block who was to win a Nobel Prize of his own), Bohr then set up an underground railway to spirit Jews out of Denmark. After the war, he began a tireless campaign against further deployment of nuclear weapons.

Midway through his scientific career, Bohr made a fascinating intellectual conversion. As a young man, he was convinced that the world around him had a logical order, that natural phenomena could be explained with rigorous logic and reason. He fell away from the church because of his feeling that its doctrines were logically untenable. Later, though, as he discovered parts of the world where logic evidently did not govern, he accepted a somewhat muddier picture. By the mid-1920s, he had developed a "Principle of Complementarity," which held that physics was large enough to contain some seeming illogic and contradictions. Late in life he designed a personal coat of arms that carried the yin/yang symbol and the motto *Contraria sunt complementa.*

Bohr took his Ph.D. at Copenhagen in 1911, writing his dissertation on the still new concept of the electron, and then headed off to Cambridge to study under J. J. Thomson himself. There he became familiar with Thomson's conception of the structure of the atom—a theory known as the "raisin cake atom" because it posited a sort of sponge cake with electrons scattered about here and there like raisins. Next Bohr went to Manchester to work with another great physicist, Ernest Rutherford; there he learned of Rutherford's "nuclear atom," which posited a small nucleus set inside an amorphous cloud of electrons. Then, in 1913, Bohr set down his own picture of the atom—a hypothesis that has prevailed, with regular refinements, ever since.

The "Bohr atom" is the atom that most adults today saw in their high school science books: the "solar system" model, with electrons swirling in concentric orbits around a central nucleus. "In this picture," Bohr explained in his Nobel Prize address in 1922, "we see a striking resemblance to a planetary system, such as we have in our solar system." The key point of the Bohr picture, though, was his insistence that electrons could not orbit in just any old spot. Using quantum mechanics and some mind-boggling mathematics, Bohr determined precisely how far from the nucleus each orbit should be, and how many electrons can reside in each orbit.

Under these rules, the orbit that is farthest from the nucleus can have from one to eight electrons. The electrical characteristics of each element are determined by this outermost orbit—specifically, by the number of electrons in the outer orbit. If an atom has only one electron in the farthest orbit, that electron will not be tightly bound to the nucleus; it could break away easily. But in an atom with a full house—eight electrons—in the outer orbit, the electrons will be held tightly in place.

Working from this theory of the atom, quantum physicists could predict which materials would be good conductors of electric current. A substance that easily released electrons would supply the free-flowing electrons that make up electric current; such a substance should be a good conductor. In a material that did not release free electrons, current would not flow; it would be an insulator. Theoretically, then, the conductivity of any material would be determined by the number of electrons in its outermost ring.

No one has ever seen an atom. Until we do, parts of the quantum picture of atomic structure will remain, in a strict sense, merely theory. The quantum view of conductivity, however, can be tested, because of the experiments that determined the conductivity of specific materials. When these experimental results are compared with the predictions of quantum theory, theory and experiment match perfectly. ,

The materials found to be the best conductors—silver, copper, gold—are indeed elements with a single electron in the outermost orbit. Materials that have proven the best insulators are indeed those with eight outer electrons. As a general matter, elements with three or fewer outer-ring electrons are conductors, and those with five or more are insulators. At the precise center of this continuum stand the semiconductors. Semiconductors, such as silicon and germanium, have four electrons in the outermost ring.

It is this special feature of semiconductor materials that makes them so spectacularly useful in electronics. Sitting on the fence, perched midway between the conductors and the insulators, semiconductors can perform valuable electronic service precisely because of their in-between structure.

Because semiconductors are right on the borderline between conductors and insulators, experimenters have found ways to alter their conductivity. This is done by a process called doping. Here's how it works: In a solid block of pure silicon, the individual atoms tend to link up with their neighbors. Each atom has four outer-ring electrons; they form a tight four-corner connection with the four outermost elements of the atom next door. When that happens, all eight electrons are held tightly in place; no electrons can break free, and no current flows. But humans have learned how to fool the silicon atoms by doping the silicon with impurities. They do it by introducing tiny quantities of a different element—arsenic, for example—that has five outer-ring electrons. The four outside electrons of a silicon atom will bind themselves to four of the arsenic electrons, leaving one extra arsenic atom unbound, free to move. Thanks to the arsenic doping, the block of silicon is now a conductor. If more arsenic atoms are introduced, more free electrons result and more current will flow. The current moving through the silicon—a flow of free electrons—is the same current that J. J. Thomson saw moving through the vacuum in his cathode ray tube.

But what if the silicon is doped with an element—boron, for example—that has only three electrons in its outermost ring? When silicon atoms try to link up with boron, the arrangement comes up one electron short, leaving a vacant spot—an empty hole—where the eighth electron should be. A nearby electron will be pulled over to fill the hole; this leaves another hole where the electron came from. Another electron moves in to fill this new hole, leaving another hole in its place. The result, effectively, is a movement of holes across the silicon block.

In his classic text on semiconductor physics, William B. Shockley explains this by comparing a block of silicon to a parking lot. In a pure, undoped crystal, every space on the lot is filled and no traffic can flow. If one car is removed, leaving a vacant slot, another car can move ahead, leaving its place open, in turn, for another car to move into. It is the cars that move, of course, but to an observer looking down every once in a while from a high building it appears that the empty space is migrating across the lot. Effectively, Shockley's book says, "the vacant parking place . . . can move owing to the successive motion of vehicles into it."

The doping process results in two different types of silicon. Where the silicon has received extra electrons, it takes on a negative charge, because electrons are negative. Where the silicon is pocked with holes—representing missing electrons—it takes on a net positive charge.

If the phenomenon of holes had been discovered in Thomson's day, when all men of science had a firm grounding in the classics, the vacant spot would most likely have been given a Greek or Latin name, a name like "vacutron" or "nihilon" or some such. The flow of holes was not firmly established, though, until the 1930s, when the classics were in decline and English had become the lingua franca of physics. Consequently, the formal scientific name for the positively charged hole is a simple English word: "hole." (Shockley titled his definitive text *Electrons and Holes in Semiconductors.*) With equivalent simplicity, physicists decided to refer to semiconductor material that had been doped with excess electrons—and is thus negatively charged—as an N-type semiconductor. A semiconductor block doped with positive charge— holes—is called P-type.

By the late 1930s, when all these intricacies had been worked out, physicists had a reasonably decent picture of what goes on inside a semiconductor. All that remained to put this knowledge to practical use—to launch the semiconductor revolution—was human ingenuity. This essential ingredient was to come from a team of three Americans headed by an intriguing figure who has been, in different seasons, one of the most respected and most reviled of all modern scientists, William B. Shockley.

Shockley was the only child of a technology-oriented couple; his father was a mining engineer, his mother a geologist. Born in London, where his father was stationed, in 1910, he grew up at the northern edge of what is now Silicon Valley. After graduating from Cal Tech he did graduate work at MIT, focusing on the movement of electrons in solid materials. Fresh out of school, the

young Ph.D. went to work for Bell Labs in 1936. Management, ignoring his graduate work, shunted him off to the vacuum tube department; by sheer persistence, Shockley eventually worked his
way into the semiconductor laboratory.

It was a rich, exciting time to be there: semiconductor physics
was just falling into place, and there was a sense among the
bright, ambitious young men in the field that they were witness-
ing the dawn of marvelous things. Buffeted by these currents of
the new, Shockley one day had an amazing idea—a Nobel
Prize–worthy idea, as it turned out. It was December 29, 1939,
and he wrote it down right away in his lab notebook: "It has today
occurred to me that an amplifier using semiconductors rather
than vacuum is in principle possible." Within a decade, he had
turned that principle into practice in the form of the transistor, an
invention that won Shockley and his colleagues, Walter Brattain
and John Bardeen, the Nobel Prize in 1956. Thereafter, with the
apotheosis of science that followed *Sputnik*, the mass media made
him into a national hero.

In addition to his work in theoretical physics, Shockley was involved in engineering (he earned more than ninety patents),
teaching, and strategic planning (he devised submarine attack
methodologies during World War II). His experience in these var-
ied intellectual disciplines prompted him to do a great deal of
thinking about thinking—about the motivations and the thought
processes that lead to good ideas. His basic rule for solving any
problem was to go back to fundamentals: "Try simplest cases."

Understanding is most likely to result, Shockley taught, from
reducing the situation to its simplest elements and proceeding
from there. His famous book on semiconductors follows the pattern: In Chapter 1 he sets forth the comparison of atomic structures to parking lots, complete with little sketches of cars in a lot
to demonstrate traffic flow. By Chapter 15 he is explaining that
"the wave function A (φ) for the hole-wave packet is not an eigen-

function for the Hamiltonian for the 2N-1 electrons in the va-
lence band."

Motivation is at least as important as method for the serious
thinker, Shockley believed, and his scientific papers are replete
with asides about the role of this comment or that experiment in
spurring him on to new discoveries. And he always maintained
that the crucial motivational issue—in fact, the essential element
for successful work in any field—was "the will to think." This was
a phrase he learned from the nuclear physicist Enrico Fermi in
1940, and never forgot. "In these four words," Shockley wrote
later, "[Fermi] distilled the essence of a very significant insight: A
competent thinker will be reluctant to commit himself to the effort that tedious and precise thinking demands—he will lack 'the
will to think'—unless he has the conviction that something
worthwhile will be done with the results of his efforts." The discipline of competent thinking is important throughout life, Shockley says, whether on a prize-winning experiment or a pop quiz in
freshman physics. For many years at Stanford he taught a freshman seminar called "Mental Tools for Scientific Thinking." The
basic text was the professor's own essay, "THINKING about
THINKING improves THINKING."

All this THINKING about THINKING thrust Shockley into a
furious racial and political controversy that he initially provoked
in the late 1960s and that continues, at lower intensity, to this day.
Perhaps it was the self-confidence that came from the knowledge
that he had hit on one of the most important scientific ideas of the
century; perhaps it was some political inclination. For whatever
reason, though, the great physicist decided that he was also an expert in the field of human intelligence. In a letter to the National
Academy of Sciences, and then in a series of lectures and interviews, Shockley urged detailed study of a problem he named "dysgenics" and defined as "retrogressive evolution through the
disproportionate reproduction of the genetically disabled." In

plain English, Shockley was claiming that, on the average, blacks are dumber than whites. Thus high birthrates among blacks could lead to "decline of our nation's human quality." "My research leads me inescapably to the opinion," Shockley said, "that the major cause of American Negroes' intellectual and social deficits is hereditary and is racially genetic in origin." Shockley went on to propose a social policy to deal with this situation: government should work to reduce birthrates among low-IQ elements of the population, through programs such as tax breaks for voluntary sterilization.

This theory, proposed by a physicist with no training in genetics and set forth during a fairly tumultuous period in American history, made Shockley one of the most despised and vilified men in the United States. He was denounced as a pseudoscientist, a fanatic, a fascist. He was burned in effigy on both coasts and denied the right to speak at some of the nation's most prestigious colleges, including Harvard, Dartmouth, and Yale. He was permitted to appear at Princeton, speaking in a small hall (chosen for its tight security) while hundreds of enraged demonstrators screamed protests outside. Back at Stanford, groups of students equipped with bullhorns, the portable and immensely powerful loudspeakers made possible by the transistor, would gather under his office window to chant "Off Pig Shockley"—affording Shockley the experience, probably unique in engineering history, of watching his own invention used to provide hundredfold amplification of demands for his death.

The instigator seemed to relish the stir he had created. A photograph of one of Shockley's classes that has been disrupted by demonstrators in Ku Klux Klan robes and pointed hats shows the professor standing calmly aside, arms folded, taking in the scene with an appearance of benign unconcern. At one of the countless "Off Shockley" rallies at Stanford, the microphone went on the blink; the inventor, who was present, stepped forward and repaired the transistorized device so that the speakers could continue their denunciation of his ideas. No matter how bitter the attacks on him, Shockley never stopped seeking new forums to convey his message; in 1982, to the dismay of party leaders, he briefly pursued the Republican nomination for the U.S. Senate. His platform called for a public inquiry into "dysgenics." Perhaps the saddest aspect of the whole thing was that the argument about intelligence became Shockley's central concern; the sheer pleasure he had always taken in pondering the mysteries of the physical world was overwhelmed by his immersion in politics. When I pushed and pulled at him to get him to remember the thought processes that led to his greatest invention, he kept steering the conversation back to intelligence and race. He was still fixated on that issue when he died in 1989.

One of the ironies of this extended controversy was that Shockley, for all his intensity and determination, seemed much more a pleasant grandfatherly type than a public ogre. Those who worked with him over the years invariably described him as "charming"; at scientific meetings he was known for telling jokes and even performing magic tricks at the podium while delivering his papers. He also had a reputation for getting the most out of the people who worked with him. This he certainly did at Bell Labs when, just after World War II, he was put in charge of a team investigating new semiconductor applications.

The senior member of the group, Walter Houser Brattain, was forty-five when the transistor was invented. Brattain grew up on the family ranch in Washington State and went to Whitman College. He did graduate work at Oregon and Minnesota, and upon receiving his Ph.D. in 1929 went to work for the newly organized research institution Bell Labs. It was a time, as he recalled later, when "the vacuum tube and thermionics were just shedding their baby teeth," and he was first put to work on tubes. Later he moved into semiconductor research under the great scholar C. J. Davisson and was present on the day in 1937 when word arrived that Davisson had won the Nobel Prize. A horde of reporters swarmed onto the premises, and the lab was quickly engulfed in microphones, newsreel cameras, and banks of klieg lights. Davisson, noticing the astonishment on his young assistant's face, stepped over to Brattain and whispered, "Don't worry, Walter—you'll win one some day."

The most unassuming of men, Brattain seemed almost embarrassed about winning awards. At the ceremony during which King Gustav VI of Sweden presented him the Nobel Prize, Brattain observed that "one indeed needs to be humble about accepting such an award when he thinks how fortunate he was to be in the right environment at the right time." He was equally diffident about the scientist's role in society. His job, Brattain said, was merely to enhance our understanding of the physical world; "I feel strongly, however, that the scientist has no right to dictate how his understanding is used." Nonetheless, when he took a look back on the twenty-fifth anniversary of the transistor, Brattain did voice one complaint: "The thing I deplore most is the use of solid state electronics by rock and roll musicians to raise the level of sound to where it is both painful and injurious."

Brattain was the experimentalist of the transistor trio; he had an intuitive feeling for the way semiconductors ought to work, and it was he who turned his colleagues' theories into working apparatus. The first transistor was an ungainly construct of germanium and wire that could be made to work only by Walter Brattain and only, he wrote later, "if I wiggled it just right."

The third member of the transistor team was John Bardeen. Born in 1908, the son of the dean of the medical school at the University of Wisconsin, he grew up in Madison and took his engineering degree at Wisconsin. He went to work for Gulf Oil in Pittsburgh but decided after three years that he preferred pure to applied science. He enrolled at Princeton, studied semiconductor physics under the future Nobel laureate Eugene Wigner, and took a Ph.D. in 1936. He taught at Minnesota, worked in the Naval Ordnance Laboratory during the war, and then agreed to join his friend Shockley at Bell Labs in 1946. In 1951, with the basic transistor work completed, he suggested that Bell undertake work on superconductors—that is, materials that show almost no electrical resistance at extremely low temperatures. Bell rejected this proposal, foolishly, on the grounds that superconductivity was too pie-in-the-sky to work. Bardeen left the lab and took a position at the University of Illinois (where a young engineer named Jack Kilby heard him lecture). While there, Bardeen joined with two other physicists, Leon Cooper and John Schrieffer, to develop the theoretical basis of superconductivity. Their scheme, known as the BCS theory, has turned out to work and has been the basis for almost all superconductor development. For this theory, Bardeen won a second Nobel Prize.

Bardeen was a perfect complement to Brattain. His forte is theory. On the transistor project, Brattain would work his experiments and then Bardeen would sift through the results and explain what they meant. He was not the type of person who could wiggle a piece of machinery just right and make it work. On the day in 1972 when his second Nobel was announced, the public relations people at Illinois, proud as punch, arranged a press conference and summoned every science reporter within traveling distance. The reporters showed up, but not Bardeen. He apologized later, explaining that he had been stuck at home because he couldn't get his (transistorized) garage door opener to work.

Brattain the experimentalist, Bardeen the theorist, and Shockley, who did some of both and was clearly the leader of the group, began in 1946 to study the different attributes of the two types of semiconductor: P-type, the material that had been doped with excess positive charges (holes), and N-type, which had excess negative charges (electrons). They worked with "bipolar" germanium—that is, a strip of germanium that had been doped so as to be N-type on one end and P-type at the other. The men focused intensively on the point in the center where N-type and P-type meet. This point, the most important meeting place in modern physics, is called the P-N junction. The P-N junction makes wonderful things happen.

The P-N junction works like the turnstile you pass through when you enter the subway or a stadium: you can go through easily in one direction but not the other. The P-N junction is a one-way door for electrons; they can pass through it going one direction, but not the other. When the semiconductor strip is hooked up to a source of current—a battery, for example—electrons can flow easily from the N-type material, across the P-N junction, to the P end. But they can't cross the junction in the other direction.

A device that lets current pass in only one direction—that's a rectifier, just like John A. Fleming's vacuum tube rectifier, which made reliable radios possible. Since the two ends of the bipolar strip act like the two electrodes of Fleming's tube, the semiconductor rectifier, like the tube, is called a diode. But instead of the large, hot, fragile, power-hungry vacuum tube that Fleming used, this semiconductor diode, with its minute P-N junction, is tiny, low-power, unheated, and unbreakable. The semiconductor diode needs no vacuum, either, and thus no glass bulb. The electronic action takes place within the solid; hence the term "solid state."

By 1946, when the Bell Labs team began its work, the operation of the semiconductor diode was fairly well understood. Semiconductor diodes were not at all important in electronics, however—except for highly specialized applications, like radar—because they did not perform the essential task of amplification. As long as electronic equipment still needed vacuum tubes—with all their attendant problems of high power, heat, and size—as am-

plifiers, it was just as convenient to use vacuum tube diode rectifiers as well. The revolution in electronics came when Shockley,
Bardeen, and Brattain, after two years of intense and frequently
exasperating work, devised a semiconductor triode—a tiny, low-
power, unheated, solid-state amplifier. This was the very device
Shockley had thought of back in December 1939.

The world's first solid-state amplifier was a jury-rigged affair put
together by Bardeen and Brattain. It involved two fine wires
placed extremely close to each other precisely at the P-N junction
on a small piece of germanium. If the wires were set at exactly the
right spot, and if Brattain wiggled them just right, a small current
flowing into one wire could be amplified to a current one hundred times as great.

The inventors demonstrated this device to the Bell Labs brass
on December 23, 1947. At first they measured the amplification
with standard electronic meters so that the observers could see the
dial on the meter swing far to the right, showing a huge amplification of current. But the test that really brought home what had
been achieved came when they hooked up a microphone to one
end of their invention and a loudspeaker to the other. One by one,
the men picked up the microphone and whispered "Hello"; the
loudspeaker at the other end of the circuit shouted "HELLO!"
Shockley, with an acute sense of history, realized the nice sym-
metry of the moment: another major breakthrough in communications had occurred in Bell's lab. "Hearing speech amplified by
the transistor," he wrote later, "was in the tradition of Alexander
Graham Bell's famous 'Mr. Watson, come here, I want you.' "

This first device, known as a point-contact transistor, was a
cumbersome, imprecise instrument that of itself would probably
not have made a significant impact. The true importance of that
first transistor was that it inspired Shockley to perfect an improved
solid-state amplifier, the device that would revolutionize electronics: the junction transistor. Shockley had only a small role in

the development of the point-contact invention. "My elation with the group's success was tempered . . . ," he recalled later. "I experienced some frustration that my personal efforts . . . had not resulted in a significant inventive contribution of my own." This frustration, coupled with the knowledge that solid-state amplification was indeed possible, gave Shockley "the will to think." For the next five weeks, he dedicated himself to "the effort that tedious and precise thinking demands." Working alone on New Year's Eve, he filled nineteen pages in his lab notebook with ideas and sketches. On January 23, 1948, he set down in his notebook the concept of the junction transistor.

Shockley's idea—the basis of all transistors today—was a semiconductor sandwich, a strip of germanium with three different regions—N-type at one end, P-type in the middle, N-type at the other end. This created two P-N junctions, back-to-back. Three wires were hooked up—one to each N-type region, and one to the P-type in the middle. Now, if a charge was sent to the center region, it would vigorously suck electrons from one N-region, across the center, and out the other end. The stronger the charge on the center region, the more electrons it drove across the junction. That is, the charge in the center section could amplify the flow of current down the strip.

In one common type of transistor, the first N-type region is called the source; the center region is the gate; the second N-type region the drain. Current flows out of the source and down the drain. How much current flows is controlled by the gate in the middle. A small change in the current hooked to the gate causes a big change in the current from source to drain. More important, variations in the gate current are exactly mimicked in the current flowing to the drain.

The three sections of this solid-state sandwich are analogous to the three electrodes of Lee De Forest's vacuum triode. The semiconductor "gate" acts like the vacuum tube "grid." A radio signal sent to the gate can shape a much stronger current flowing from

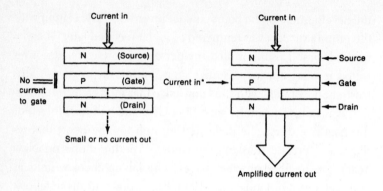

source to drain. Just as in the vacuum tube, the effective result is an amplified reproduction of the weak signal. In addition to its application as an amplifier, the solid-state triode can serve as an extremely high-speed switch. If the device is adjusted properly, a signal to the gate will cut off the drain current completely; another signal will open the gate and turn the drain current back on. The important thing is that, in a transistor, this on-off switching takes place in billionths of a second. This is faster than any vacuum tube could be made to switch. This capability for ultra-high-speed switching made possible the modern computer and all its myriad offspring and networks.

Smaller, lighter, faster, more sensitive, more reliable, and far more power-efficient than the vacuum tube, the transistor could not have been anything other than a stupendous success. Two important decisions by the Bell System accelerated its progress. Mindful of its founder's lifelong interest in helping the deaf, Bell waived all patent royalties on the first important transistor product, the miniature hearing aid. For all other applications, Bell Labs, moved by a sense of public service (and, perhaps, by a pending antitrust action against AT&T, its parent company), established a bargain-basement license fee of $25,000 and ran training programs for all firms interested in producing transistors.

By the mid-fifties, semiconductor sales were in the billion-dollar range and the vacuum triode was becoming a museum piece. Each new year brought hundreds of intriguing new inventions based on the transistor. The popular press treated the new technology as a full-fledged miracle. "To all industrial needs, and most human physical needs," *Time* magazine noted in a typically breathless report in 1957, "the electronics magicians are sure they have the key."

Or were they? By 1957, the electronic magicians were sure only that they had a serious problem—the problem posed by the tyranny of numbers. Unless the interconnections dilemma could be resolved, the enormous promise of a transistorized future might never be realized. The central importance of the problem—and the profits to be gained from its solution—instilled in governments and research labs and manufacturing concerns around the globe "the will to think" about a solution, and the will to spend large sums in pursuit of it.

One of the firms in the forefront of this pursuit was Texas Instruments, which mounted a large-scale research project to deal with the numbers problem early in 1958 and began recruiting semiconductor experts from throughout the world for the task. Among the men TI hired was a lanky thirty-four-year-old engineer from Milwaukee named Jack Kilby.

A NONOBVIOUS SOLUTION

Jack St. Clair Kilby landed his first job in electronics the year the transistor was born and has been working ever since at the front lines of high technology. He was among a small group of pioneers in the 1950s who developed the transistor from a laboratory specimen to a mainstay of industry. He conceived and built the world's first integrated circuit, or semiconductor chip, and then proceeded to invent what is probably the chip's most famous offspring, the handheld calculator. He spent a half dozen years after that working on the photovoltaic effect, an as yet imperfectly understood phenomenon of semiconductor physics, in an effort to construct an electric generating station fueled by the sun. He has won countless scholarly awards, including the National Medal of Science, the technologist's equivalent of the Congressional Medal of Honor; the Kyoto Prize, Japan's equivalent of the Nobel Prize; and the Nobel Prize itself. His picture hangs in a hall of fame in Washington between those of Henry Ford and Ernest Lawrence, inventor of the cyclotron.

For all that, however, Jack Kilby in person seems the very antithesis of high tech. Calm and quiet, plain-spoken and plainly dressed (his standard uniform in the lab is an open-necked sports shirt and knockabout cotton trousers), Kilby is the kind of person you might expect to find rocking peacefully on the porch of some country store, his large feet propped up on the railing. He is an imposing figure, not fat but big in every other sense: six feet six inches tall, wide shoulders, massive hands, a large, round, ruddy face framed by a few wayward tufts of gray hair poking up from the temples, and an enormous smile that suggests, accurately, a friendly, casual, unruffled personality. An introvert, he spends a good deal of time alone with his thoughts, working through ideas; he has always done his most creative work on his own. In conversation, he is not quick. Ask him a question—about semiconductors, politics, the best route to the airport—and he will take a long puff on a Carlton, think for a moment in absolute silence, take another puff, and then answer, softly, slowly. The answers are invariably thoughtful and delivered in fully structured sentences that flow perfectly from beginning to end without digression or detour.

Despite his pioneering work in the most modern of technologies, the inventor has an old-fashioned streak. He won't wear a digital watch; characteristically, he has given considerable thought to the difference between digital and analog (i.e., with hands) timepieces and concluded that the older kind better conveys the seamless passage of time. He was one of the last people in the high-tech community in Dallas to get a personal computer, and has never dreamed of doing anything more complicated with the machine than trying, not always successfully, to attach a document to an e-mail. When Kilby was called upon in December 2000 to give the Nobel lecture in physics, functionaries from the Royal Swedish Academy contacted him in advance to find out which software program he would be using to prepare the presentations for his talk. Software? Presentations? "Gosh, I guess I'll just have a couple of slides," Jack said. Although he is probably the single person most responsible for the demise of the slide rule, he still keeps his favorite Keuffel & Esser Log-Log Decitrig handy in the center drawer of his desk, and in some ways he prefers it to the handheld calculators that rendered it obsolete. "It's an elegant tool," he says affectionately. His hobby is photography (black-and-white, of course), for which he has contentedly used the same trusty Hasselblad for thirty years. He tends to keep his cars well past the 100,000-mile mark. "As long as it runs, what the heck," he explains.

In an industry and a company (he has worked at Texas Instruments on and off for almost half a century) where "tough" and "aggressive" are terms of high praise, the quiet but friendly inventor is famous as a nice guy. As he treks through the meandering hallways of Texas Instruments' Dallas headquarters—walking with the wary, stooped gait of a man who has bumped his head too often on low ceilings—he greets everybody by name, from top management to messenger boys. Everyone at the company seems to have a story or three about some act of kindness on Kilby's part. To that collection I can add another. When I first called Kilby, out of the blue, to ask if I might spend some time with him, he readily agreed, and then added—it was a first in my journalistic career—that he would pick me up at the airport and shuttle me around because "taxis can be hard as hell to get around here."

Because of the more-or-less parallel development of quantum theory and semiconductor technology, Kilby's work has regularly taken him near or right up to the leading edge of physics. It is a bewilderingly complex field; to understand the flow of charge in a chip, for example, one has to calculate the eigenvalues of the z-component of the angular momentum operator. But Kilby is not a scientist. He is quite firm on the point, insisting, in soft but definite tones, that he is an engineer. "There's a pretty key difference," he says. "A scientist is motivated by knowledge; he basically wants to explain something. An engineer's drive is to solve problems, to make something work. . . . That is basically what I have always wanted to do, to solve technical problems. It is quite satisfying, extremely satisfying, to go through the process and find a solution that works."

Kilby has done a great deal of thinking about that process, and, true to form, he has worked out a careful theory of the art of solving a problem, technical or otherwise. Somewhat simplified, the method involves two levels of concentrated thought.

At first, the problem solver has to look things over with a wide-angle lens, hunting down every fact that might conceivably be related to some kind of solution. This involves extensive reading, including all the obvious technical literature but also a broad range of other publications—books, broadsides, newspapers, magazines, speeches, catalogues, whatever happens into view. Kilby himself reads, not skims but reads, two or three newspapers every day and a dozen magazines or so each week. In addition he devours books; his office looks like a publisher's warehouse where the books have staged a coup. For years, Kilby took the time to read every new patent issued by the U.S. government. Some of the inventions were arguably related to his work: "anode stud coatings for electrolytic cells"; "electric current regulator." Others seemed rather far removed: "gas-fired ceramic radiant poultry brooder"; "dentifrice encapsulation"; "guitar amplifying system"; "self-packaged glider toy"; "golf putting aid"; "3- or 4-product surface-wave acousto-optic time-integrating correlator." "That's all right," Kilby says. "You read everything—that's part of the job. You accumulate all this trivia, and you hope that someday maybe a millionth of it will be useful." For recreation, Kilby says, "I read trash."

The next step in Kilby's system requires switching to an extremely narrow focus, thinking strictly about the problem and tuning out the rest of the world. This requires, first of all, an accurate definition of the problem. "The definition of the problem becomes a major part of the innovation," Kilby has written. "A lot of solutions fail," he says, "because they're solving the wrong problem, and nobody realizes that until the patent is filed and they've built the thing." It is also necessary to develop a clear understanding of the natural constraints surrounding the problem; the heart of the inventor's job is finding a way to slip past the roadblocks erected by nature. "Although invention is considered a creative process," Kilby said once in a lecture on the subject, "it differs appreciably from creativity in the arts. The painter starts with a blank canvas, the author or poet with a blank sheet of paper. They choose an image . . . and they are free to use any techniques they have to achieve it. Technical creativity is more constrained. The laws of nature, the properties of materials . . . provide very real constraints."

In this concentrated, single-minded focus on the question at hand, the problem solver must also tune out all the obvious solutions. This is a key principle, important to emphasize because it is somewhat counterintuitive. The mind tends to jump to the answer that is immediately evident. In fact, this answer is probably wrong. If the problem is of any importance, all the obvious solutions have been tried already. The word "nonobvious" appears in few dictionaries, but it is an important part of Kilby's personal lexicon, a concept he returns to again and again when he gets talking about the business of solving problems. Some of history's most important innovations, he says, were so nonobvious as to violate the scientific rules of the day. "You only arrived at the invention when somebody developed a method that everyone else had already decided was obviously wrong."

At this point, if the problem solver's preparation has been broad enough, and he has defined the right problem, and he observes the physical limits, and he's creative, and he's lucky, he might hit on a nonobvious solution that works. But that is not enough, at least not for an engineering problem. The essence of engineering,

Kilby says, is cost consciousness. "You could design a nuclear-powered baby bottle warmer," he says, "and it might work, but it's not an engineering solution. It won't make sense in terms of cost. The way my dad always liked to put it was that an engineer could find a way to do for one dollar what everybody else could do for two."

Kilby's dad was an electrical engineer who worked at electric power companies around the Midwest and eventually rose to the presidency of the Kansas Power Company, a medium-size utility that had small generating plants scattered around the western part of the state and headquarters in Great Bend, a neat, bustling town at the point where the Arkansas River bends south toward the Mississippi. Jack was born in Jefferson City, Missouri, in 1923, but spent most of his childhood in Great Bend. He attended the public schools there, and in the summers he and his father would traverse the plains in a big 1935 Buick, stopping at each of the power company's remote facilities. They would crawl through the works of the generating stations, trying to find out what had gone wrong with a faulty armature or testing the efficiency of a new-model transformer.

When the blizzard of '37 swept through Kansas, blocking roads and felling telephone lines everywhere, the senior Kilby borrowed a neighbor's ham radio to keep track of his customers and his far-flung operations. Naturally, the curious son had to see how this gadget worked. Twisting the dials, turning the big antenna this way and that, squeezing the earphones tight against his head so that he could make out the weak, fluctuating signals racing through the Kansas night, Jack was quickly hooked on radio. Partly it was the sheer fascination of the tool itself, and curiosity about why it worked. But the teenager could also appreciate how the power of human ingenuity had improved daily life for ordinary people. "It was during an ice storm in my teens," he recalled some sixty years later, "that I first saw how radio and, by extension,

electronics, could really impact people's lives by keeping them in-
formed and connected, and giving them hope." A brand-new government agency, the Federal Communications Commission,
began requiring licenses for radio operators. Jack studied for
weeks, took the exam, and came home from school one day to
find an official letter assigning him his own set of call letters:
W9GTY. He built a ham set, improved it, scavenged some parts,
improved it again. By the time he got to Great Bend High School
it was clear that he would make his career in electrical engineering, and he set his sights on the engineer's mecca, the Massachusetts Institute of Technology. On a June day in 1941 he boarded
the *New England States,* the crack train connecting the midwest-
ern plains with the great centers of learning and commerce on the
East Coast, and rode to Cambridge. He spent a month there training to take the entrance exam for MIT.

At this point, our inventor's story takes a downward turn. Jack
failed the test. He was turned down by MIT. Decades later, having launched the Second Industrial Revolution, received more
than fifty patents, and won all the leading engineering awards,
Kilby still felt the sting of flunking that exam. He could remember his score on the test—he got 497, three points short of passing.
Of course, it all worked out for the best in the long run, as Jack
pointed out on the day he won the Nobel Prize. But at the time
that rejection letter from MIT was devastating. Moreover, it created a practical crisis. Jack had not bothered to apply to any other
college. After some scrambling he was admitted to his parents'
alma mater, the University of Illinois. He had been there less than
four months when the Japanese bombed Pearl Harbor. Freshman
Kilby became Corporal Kilby, assigned to a radio repair shop at an
Army outpost on a tea plantation in northeastern India.

Everybody learns something in the Army. The eighteen-year-
old corporal learned that creative engineering can solve problems
that have been officially declared insoluble. His unit was one of

the first endeavors of the United States in guerrilla warfare. Small teams of soldiers would be airlifted "over the hump"—that is, across the Himalayas—into Burma to put together indigenous resistance movements to harass the Japanese. The Americans kept in touch with their base using radios they carried on their backs. The best "portable" radios ever made were provided, but they weighed 60 pounds and broke down regularly under the stress of jungle operations. The Army responded to all complaints by saying that its transmitters, the state of the art in radio, could not be improved. An engineer, of course, knows that there is no machine anywhere that cannot be improved. The engineers in Kilby's unit set up a lab in a dusty pup tent. They sent Corporal Kilby down to Calcutta to buy old radio parts on the black market. Over time, they began turning out ad hoc transmitters that were both lighter and more power-efficient than the official issue.

After the war, Kilby went back to Illinois, eager to learn about radar and other wartime advances in electronics. On the whole, he was disappointed. The one thing he can recall now about his electronics classes was that none of the experiments turned out the way the instructors said they would. Further, the smattering of physics classes offered to engineering students didn't cover the crucial twentieth-century developments revealing the nature of subatomic architectures. There were courses at Illinois on quantum physics and semiconductor phenomena, but they were restricted to scientists; "they weren't going to expose that funny stuff to simpleminded engineers," Jack says. Kilby graduated in 1947 with a traditional engineering education and decent but not outstanding grades. He went to work in Milwaukee at Centralab, for the excellent reason that it was the only firm in his field that offered him a job.

As such things often do, the job turned out to be a perfect spot for the neophyte engineer. Centralab, a division of Globe-Union Corporation, was then producing electric parts for hearing-aid,

radio, and television circuits. It was an intensely competitive business, where a cost differential of one dollar per thousand parts—a
tenth of a penny per part—could win or lose huge contracts. "It
was sort of a crash course in sensitivity to cost," Kilby recalled
later. By the late 1940s, radio engineers had already determined
the optimum material for each kind of electric component; resistors were made of carbon, capacitors of metal and porcelain, connecting wires of silver or copper. Rather than try to squeeze out a
few additional hundredths of a cent per part by improving materials, Centralab was working on the notion that more could be
saved by production efficiencies—by placing all the parts of a circuit on a single ceramic base in one manufacturing operation.
The firm had only mixed success, but the conceptual seed—that
the components of a circuit need not be manufactured separately—was to stay with Kilby and bear important fruit.

In his early years at Centralab, Kilby began to develop his problem-solving methodology. To acquire the wide-angle picture of
the problems he had to solve at work, he was determined to learn
everything he could about his field. He took graduate courses at
night, plowed through the technical literature, attended any lecture that might be interesting. One night at Marquette University
he heard a physicist—it was John Bardeen—describe a new invention that achieved amplification and rapid on-off switching
without a vacuum tube. Amazing! The very existence of such a
device seemed to challenge all the rules of electric circuits that
Jack had learned in school. Kilby set out to read everything he
could find about this new solid-state device. Then as now, Kilby's
reading went far beyond electronics. At one point he happened
upon a dental supply catalogue; one page, so unpleasant he can
still remember it, described a new technique that used small sandblasters to scour away tooth decay.

Centralab was small and informal enough to let the most junior man in the lab take on important jobs. Kilby was immediately

put to work solving real engineering problems, and as he got the hang of it, he acquired a priceless asset for anyone engaged in creative work: confidence. Over time, Jack came to realize that if he approached a problem correctly, worked at it long enough, and refused to let initial failures get him down, he could find a solution. One of his first real successes, in fact, solved one of the major problems with Centralab's single-step circuit-building process—the reliability of the resistors. The process involved making resistors by printing small patches of carbon on the ceramic base. The machinery was imprecise, so no two carbon patches were the same size. That meant no two resistors performed exactly the same, which made it impossible to build a reliable circuit. Kilby was given the job of finding a way to make all the resistors the same size. It had to be done quickly, the boss said, and whatever solution Jack came up with had to be (a) cheap and (b) simple. Starting off with his wide-angle review of the problem, Kilby pondered everything that might be relevant. Something came to mind, something he had read someplace. Those tiny dental sandblasters—did anybody use them? As it happened, the technique had never caught on with dentists because their patients found it repulsive. But Kilby managed to track down some of those precise devices; sure enough, they were perfect for carving away excess carbon and making all the printed resistors the same size. It was a nonobvious approach that finally made it possible for Centralab's single-step process to produce reliable circuits.

When Bell Labs announced in 1952 that it would issue licenses for production of its newly patented transistor, Centralab put up the $25,000 license fee and dispatched Kilby to Bell's ten-day crash course in the new technology. There he got a detailed look at the fantastic new world that would be possible now that circuits would be free of the limitations imposed by vacuum tubes. He came back to Milwaukee full of ideas, ideas that led to important advances in the manufacture and packaging of transistorized equipment. Gradually, however, he came to realize that the new electronic world had a limit of its own.

Working in a relatively small firm, where the circuit designers in the engineering lab had regular contact with the plant managers, Kilby soon learned—probably sooner than many other people in the business—that realities of the manufacturing process severely restricted the complexity of transistorized circuitry. Upstairs in the lab, Kilby and his colleagues could design a hearing aid or a radio amplifier that squeezed unheard-of numbers of components into minute spaces. But down in the factory, those circuits could not be built; there were just too many interconnections too close together for the human hand to make them. "Jack Morton at Bell Labs suggested that electronics was facing a 'tyranny of numbers,'" Kilby recalled, "and that was a perfect term for it because the numbers of parts and connections in some of these new circuits were just too big. The simple fact was that you could not do everything that an engineer would want to do."

For Kilby, the recognition of this major new problem was electrifying. Just as he was coming into his own as an engineer, a problem solver, the world of electronics was up against a baffling problem of premier importance. The advent of the transistor offered enormous, world-shaking possibilities, but they would never be realized unless somebody found a way around the problem of numbers. There was a huge and growing gap between design and production. Like everyone else in the industry, Jack Kilby plunged into the search for a way to bridge that gap.

It was evident, though, that solving the tyranny of numbers, if indeed a solution could be found, was a task that would require large resources—considerably larger than a firm the size of Centralab could muster. "I felt," Kilby wrote later, ". . . that it would not be possible for very small groups with limited funding to be competitive. I decided to leave the company." Early in 1958 he sent out his résumé to engineers at a number of larger firms. Be-

cause the budding electronics industry was a thoroughly meritocratic universe, the lack of the MIT degree was irrelevant now.
Jack's reputation, and his catalogue of patents, made him a hot
commodity in the field. IBM made an offer. Motorola did as well.
And there also came a letter from Willis Adcock of Texas Instruments.

Texas Instruments today, largely because of Jack Kilby, is a
global semiconductor giant, one of the world's leading manufacturers of microelectronic devices. In 1958, though, it was just beginning to make a mark in the electronics business. The company
had been born in the mid-twenties as the Geophysical Research
Corporation; its business was sending sound waves deep into the
earth to find potential oil drilling sites. During World War II the
same deep-sounding methods proved useful for locating enemy
submarines, and GRC became a defense contractor. The firm's
postwar president, Patrick Haggerty, expanded the government
business to the point where manufacture of electronic instruments became more important than geophysical research. Convinced that great things were in store for his little firm, the
visionary Haggerty changed the company's name to General Instruments—an audacious suggestion that this impudent pup in
Dallas could stand with General Electric and the other eastern
electronics giants. The Pentagon didn't like the choice—it had
another supplier with a similar name—so Haggerty unhappily fell
back on geography: Texas Instruments.

It was another audacious Haggerty gambit that started TI on its
road to dominance. In 1952, when transistors were still exotic, unreliable devices costing $15 or more each, Haggerty hired a Bell
Labs physicist named Gordon Teal and ordered him to develop a
reliable mass-production transistor that would sell for $2.50. Teal
did it. In 1954, Haggerty launched his most famous initiative: he
put his cheap, reliable transistors into a consumer product—the
pocket radio. The idea was a smash hit in the marketplace. More

important, it made the transistor a common household item and
Texas Instruments a common name in electronics.

The first pocket radios, like all transistorized equipment of the
day, used transistors made of germanium, a material easy to work
with but unsatisfactory for many applications because it could not
operate at high temperatures. Another semiconductor material,
silicon, could withstand heat. But silicon was almost impossible
to work with; the material is brittle and difficult to purify, and
components made of silicon tended to shatter on the assembly
line. Haggerty, not the type to let impossible manufacturing ob-
stacles stand in his way, ordered Gordon Teal and Willis Adcock,
a physical chemist, to devise a silicon transistor. On the boss's
command, the project was pursued under security arrangements
that any of the world's spy agencies would admire. In the end,
Haggerty's ambition and his commitment to secrecy paid off in
spectacular fashion. In May 1954, Teal and Adcock attended a
technical meeting on manufacturing problems in the transistor
business. The holy grail at the time was a transistor made of silicon, but speaker after speaker at the meeting set forth the insu-
perable problems posed by silicon. Finally, Teal rose to speak. He
had listened with interest, he said, to the bleak predictions about
silicon's feasibility. "Our company," he noted calmly, "now has
two types of silicon transistor in production. . . . I just happen to
have some here in my coat pocket." Adcock then appeared, carrying a record player that employed a germanium transistor. As a
record played, Teal dunked the transistor in a vat of boiling oil;
the sound stopped. Next Teal wired in one of TI's new silicon
transistors. He dumped it into the hot oil; the band played on.
The meeting ended in pandemonium, for in those days long before the advent of the cellular phone, Teal's presentation sparked
a mad race to the telephone booths among salesmen and engineers from the major eastern electronics firms. A reporter from
Fortune magazine overheard a sales rep from General Electric

shouting into the phone: "They've got the silicon transistor! In Texas!!" It was not clear which of these facts was the more astounding.

Texas Instruments was on its way, and TI's triumphant engineers began producing more and more ambitious designs. Soon enough, though, the tyranny of numbers became evident to the people in Dallas. Willis Adcock was placed in charge of a major research effort to surmount this obstacle to further progress in electronics. One of the first solutions TI worked on was an idea called the Micro-Module. The theory behind it was that all the components of a circuit could be manufactured in the same size and shape, with wiring built right into each component. These identical modules could then be snapped together, like a child's Lego blocks, to make instant circuits. The concept was important to Texas Instruments, not so much because of its intrinsic merits, but because it was important to the U.S. Army. Each of the military services was pursuing its own solution to the interconnections problem, and the Army keenly desired that its proposal should prevail; if TI could deliver, it would become the darling of all Army contracting officers. Thus when Jack Kilby arrived at Adcock's lab in May 1958, the Micro-Module was the hottest thing going. Kilby disliked it from the start.

This feeling was partly an engineer's intuition; the Micro-Module bore some resemblance to an idea that had flopped at Centralab, and Kilby didn't think it would work any better in Dallas. To a systematic problem solver like Jack Kilby, though, the real flaw was more basic: the Micro-Module implied the wrong definition of the problem. The real problem posed by the tyranny of numbers was numbers. The Micro-Module did nothing to reduce the huge quantities of individual components in sophisticated circuits. No engineer could work with much enthusiasm on a solution to the wrong problem; Kilby's heart sank at the thought that he had left a good job and moved his family across the country only to be put to work on a project that was fundamentally off target.

Texas Instruments then had a mass vacation policy; everybody took off the same few weeks in July. Kilby, who hadn't been around long enough to earn vacation time, was left alone in the semiconductor lab. He was in a rotten mood. "I felt it likely," he recalled years later, "that I would be put to work on a proposal for the Micro-Module program when vacation was over—unless I came up with a good idea very quickly."

Kilby plunged in with his wide-angle approach, soaking up every fact he could about the problem at hand and the ways Texas Instruments might solve it. Among much else, he took a close, analytical look at his new firm and its operations. The obvious fact that emerged was Texas Instruments' heavy commitment to silicon. To capitalize on its victory in the race to develop silicon transistors, TI had invested millions of dollars in equipment and techniques to purify silicon and manufacture transistors with it. "If Texas Instruments was going to do something," Jack explained later, "it probably had to involve silicon." This conclusion provided Kilby the focus he needed for the narrow, concentrated phase of problem solving. He began to think, and think hard, about silicon. What could you do with silicon?

Jack Kilby's answer to that question has come to be known as the monolithic idea. The idea has so changed the world that it is just about impossible today to reconstruct what things were like before he thought of it—and thus it is almost impossible to appreciate how ingenious, and how daring, the answer was. The monolithic idea has become an elementary part of modern science, as fundamental, and as obvious, as J. J. Thomson's daring suggestion that there were tiny charged particles swirling around inside the atom. In July 1958, though, Kilby's answer was hardly elementary.

What could you do with silicon? It was already known in 1958 that the standard semiconductor devices, diodes and transistors, could be made of silicon, if the silicon was doped with the proper impurities to make it conduct electric charges. But if the silicon had no impurities, its electrons would all be bound in place. No charges would flow through such a piece of silicon; it would block current just like a standard resistor. Kilby thought about that: A silicon resistor? Why not? A strip of undoped silicon could act as a resistor. It wouldn't be as good as a standard carbon resistor, but it would work. For that matter, by taking advantage of the peculiarities of the P-N junction, you could also make a capacitor out of silicon. Not much of a capacitor, to be frank about it—its performance wouldn't equal that of a standard metal-and-porcelain capacitor—but it would work. And come to think of it—this was the idea that would revolutionize electronics—if you could make all parts of a circuit out of one material, you could manufacture all of them, all at once, in a monolithic block of that material.

The more Kilby thought about it, the more appealing this notion became. If all the parts were integrated on a single slice of silicon, you wouldn't have to wire anything together. Connections could be laid down internally within the semiconductor chip. No matter how complex the circuit was, nobody would have to solder anything together. No wires. No soldering. The numbers barrier would disappear. And without wiring or connections, an awful lot of components could be squeezed into a pretty small chip. On July 24, 1958, Kilby opened his lab notebook and wrote down the monolithic idea: "The following circuit elements could be made on a single slice: resistors, capacitor, distributed capacitor, transistor." He made rough sketches of how each of the components could be realized by proper arrangements of N-type and P-type semiconductor material.

Some four decades later, when that sentence from Jack's old notebook was read out at the Nobel Prize ceremony in Stockholm, the idea that a whole circuit could be built on a single mi-

crochip of silicon was so common it appeared in junior high textbooks. In 1958, however, this suggestion was so "nonobvious" as
to be astonishing. "Nobody would have made these components
out of semiconductor material then," Kilby has explained. "It
didn't make very good resistors or capacitors, and semiconductor
materials were considered incredibly expensive. To make a one-
cent carbon resistor from good-quality semiconductor seemed
foolish." Building a resistor out of silicon seemed about as sensible as building a boxcar out of gold; you could probably do it, but
why bother? Even Kilby was a little skeptical at first: "You couldn't
be sure that there weren't some real flaws in the scheme somewhere." The only way to find out was to build a model of this integrated circuit and give it a test. To do that, Kilby would need the
boss's okay.

When everybody came back from vacation, eager to get cracking on the Micro-Module, Kilby showed his notebook sketches to
Willis Adcock. "Willis was not as high on it as I was," Kilby recalled later. Adcock was intrigued with the idea but had doubts
about its practicality; "it was pretty damn cumbersome," he said
afterward. It was probably worth trying; but on the other hand,
Adcock was supposed to be making Micro-Modules to keep the
Army happy. To build Kilby's model, he would have to divert people from that project and put them on the previously untried task
of building a complete circuit out of semiconductors. Adcock hes-
itated. Kilby pushed. Eventually, they made a deal: if Kilby could
actually make a working resistor and a working capacitor out of
separate pieces of silicon, Adcock would authorize the far more
costly effort to construct an integrated circuit on a single semiconductor chip. Kilby painstakingly carved a resistor out of a strip
of pure silicon. Then he took a bipolar strip of silicon and wired
the P-N junction to make the capacitor. He wired these strange
devices into a test circuit, and they worked. Adcock looked it over,

and then okayed the attempt to construct a complete circuit on a single chip.

The design that Adcock chose was a phase-shift oscillator circuit, a classic unit for testing purposes because it involves all four of the standard circuit components. An oscillator is the opposite of a rectifier; it turns direct current into alternating current. If it works, the oscillator transforms a steady, direct current into fluctuating pulses of power that constantly change direction, back and forth, back and forth. The transformation shows up neatly on an oscilloscope, a piece of test equipment that displays electric currents graphically on a television screen. If you hook direct current—for example, from a battery—to the oscilloscope, the steady current will show up as a straight line across the screen, like this:

But if you put a phase-shift oscillator between the battery and the oscilloscope, the oscillating current will show up as a gracefully curving line—a sine wave—undulating across the screen, like this:

On September 12, 1958, Jack Kilby's oscillator-on-a-chip, half an inch long and narrower than a toothpick, was finally ready. Jack had glued it to a glass slide so that it would sit flat on the table; it had wires sticking out here and there. Somebody with a sense of history took a picture of the thing, and Jack suddenly felt kind of embarrassed about how crude it looked. A group of Texas Instruments executives gathered in Kilby's area in the lab to observe this tiny and wholly new species of circuit; Jack was surprised to see the chairman of the company, Mark Shepherd, among the onlookers. This thing had better work, he thought to himself. Conceptually, of course, Kilby knew it would. He had thought the thing through so often, there couldn't be a flaw. Or could there? After all, nobody had ever done anything like this before. Kilby was strangely nervous as he hooked up the wires from the battery to his small monolithic circuit, and from the circuit to the oscilloscope. He fiddled with the dials on the oscilloscope. He checked the connections. He looked up at Adcock, who gave him a here-goes-nothin' shrug. He checked the connections again. He took a deep breath. He pushed the switch. Immediately a bright green snake of light started undulating across the screen in a perfect, unending sine wave. The integrated circuit, the answer to the tyranny of numbers, had worked. The men in the room looked at the sine wave, looked at Kilby, looked at the chip, looked at the sine wave again. Then everybody broke into broad smiles. A new era in electronics had been born.

LEAP OF INSIGHT

The little airplane jumped from the boy's hand and shot off into the blue Iowa sky, the engine purring, the body spiraling perfectly, the plane racing higher and farther away every second, soaring right past the end of town and far out over the cornfields. It was almost a mile away, still performing beautifully, when he lost sight of it for good. "That was my first technological disaster," Robert Norton Noyce recalled many years later. "Yeah, it worked a lot better than any model I'd ever built before, but it was gone." Actually, it wasn't gone. Six months later, during the harvest, a farmer found the toy plane among the cornstalks and guessed that it was probably the work of the minister's son, who was always messing around with engines and gadgets and models.

By then, though, the boy had moved on to other things. That lost airplane had prompted him to build a radio control unit for his next model. Radio proved so interesting that Bob Noyce and a buddy put together a pair of crude transceivers to send messages back and forth. Neither boy obtained a radio operator's license, making their network a federal offense; the crime can safely be reported now because the statute of limitations for their violation expired about 1942. The boys got interested in chemistry; soon Bob was mixing his own home-brew explosives out of gun cotton and nitrogen tri-iodide. Then he found an old washing machine motor and tried to make it drive his bicycle. Then there was something else, and something else after that. "I was a pretty curious kid," Noyce said later, looking back. "I was always trying to figure out how everything worked."

Some things never change. The most striking thing about Robert N. Noyce—the grownup version, that is—was the enormous range of his interests and the breadth of his activities. Until the day he died, he still wanted to know how everything worked. At various stages of his professional career he was a theoretical physicist, an inventor, a corporate chief executive, a venture capitalist, a lobbyist, and eventually, elder statesman and leading spokesman for the American semiconductor industry. His colleagues and competitors called him "the mayor of Silicon Valley," and the only problem with that title was that his influence reached well beyond Silicon Valley in his last decade or so. He was successful as a scientist and engineer—he won a slew of academic and industrial awards, and surely would have shared the Nobel Prize with Jack Kilby if he had lived long enough. But he was even more successful as an entrepreneur—indeed, he set the mold for that classic turn-of-the-century phenomenon, the high-tech multimillionaire. It was Bob Noyce who first demonstrated to the denizens of Silicon Valley that a clever engineer could turn his technical talent into boundless wealth. By 1990, estimates of his net worth ranged toward the billion-dollar mark, which put him roughly in the same league as Bill Gates at the time. Off the job, he restored old airplanes and helped plan Harvard's future and skied and studied Japanese and took up, with maximum energy and commitment, whatever else his far-flung curiosity fastened upon.

An affable, naturally gregarious person, Noyce seemed to gravitate to a leadership position in most things he undertook. In the early 1980s, he got interested in madrigal singing. He studied, practiced, and joined a chorus; soon enough, he became its conductor. A friend once took Noyce to an Audubon Society meeting; soon enough, he was supervising a global effort to find a new habitat for a dwindling species of Newfoundland puffin. He was one member of a large group of Silicon Valley executives who decided a few years back that the microelectronics business needed a professional organization; soon enough, Noyce was chairman of the Semiconductor Industry Association.

A wiry, athletic-looking man with angular features and curly hair that went silvery gray in his sixties, Noyce carried himself with the rakish self-assurance of a racing driver or jet pilot (despite a clear memory of that first disaster, Noyce was always fascinated with airplanes and flew a rebuilt Republic Seabee, a World War II seaplane). A fashionable dresser, he wore silver-rimmed aviator glasses, a large gold ring, and a platinum watch that had both old-style hands and new-style digital readout. If Jack Kilby seems as loose as a spare piece of string, Noyce was a coiled spring of energy and enthusiasm. When the *Harvard Business Review* asked him to describe the employees of Intel, the massively successful microchip company he cofounded, Noyce responded with an answer that might serve as a self-portrait. "They're high achievers," he said. "High achievers love to be measured, when you really come down to it, because otherwise they can't prove to themselves that they're achieving." Among such people, Noyce added, "I don't think you could call it a relaxed atmosphere. A confident environment, but not a relaxed one." A confident, but rarely relaxed, high achiever—that description fit Bob Noyce perfectly. One of Noyce's acolytes at Intel, Andrew Grove, who took over as chairman of the firm in the 1990s, offered a somewhat more chilling description of the high-tension corporate culture: "Only the paranoid survive."

One reason it was difficult to relax at Intel when Noyce was running the place was that the firm deliberately kept itself on the leading edge of technology, always looking for the new idea. Actually, that's the way Intel still operates, more than a dozen years since the patriarch left. That's the way Robert Noyce liked it; he defined himself as a "technologist," and defined that term as "the kind of person who is comfortable with risk." And there lies the key difference between a technologist and a businessman, Noyce said: "No businessman would have developed the telephone. It's got to be a maverick—some guy who's been working with the deaf and gets the crazy idea that you could actually send the human voice over a wire. . . . A businessman would have been out taking a market survey, and since it was a nonexistent product, he would have proven conclusively that the market for a telephone was zero."

The same description—"comfortable with risk"—applied to Noyce's approach to problem solving. He attacked engineering problems with absolute confidence that there was a perfectly good solution somewhere, and that the human mind could find it. He was to be a prolific, impulsive producer of solutions himself, but he also knew that "a lot of my great ideas turn out to be ridiculous when you look closely." For that reason, Noyce was never a solitary inventor like Jack Kilby. He needed to work with others, to talk things over, to launch his ideas and see if they would fly. "You explain a lot of things to yourself," he said, "by trying to explain them to someone else. And then either you can see for yourself that the idea won't work, or the other person can spot the problem and help you find a better way."

Early in his career Noyce came into the ken of another famous technical optimist, William B. Shockley, and it is clear that he absorbed Shockley's first rule for solving problems: "Try simplest cases." Asked to describe the right way to find a solution, Noyce put it this way: "All of our progress, in any technical field—well, there are a few leaps of insight—but beyond that it has been decomposing it to its simplest elements, to understandable elements, and building it back up from its simplest element. Shockley had this wonderful ability to make the right simplifying assumption, to get the math out of the way until you had a basic visual image of what was happening. . . . Well, you see, you have to get back to the basic, the simplest picture before you can understand a problem well enough to solve it."

Like Kilby, Noyce tended to be instantly suspicious of any solution that seemed too obvious. "It's absolutely true, whether it's a technical question or anything else, that a lot of people are going to try to approach a problem the same way everybody else has," Noyce said. "You can just let yourself get in a rut. And if you don't get out of the rut, you're not going to solve the problem." The person who can jump over that rut—that person, the person who can be a successful inventor—is the one who is able to come up with the "leap of insight." "The successful solution comes about because somebody was able to fire up his imagination and try something new . . . ," Noyce added. "If you want to achieve something worthwhile, you have to jump to the new idea."

Noyce spent his entire boyhood pursuing new ideas, new phenomena, new gadgets. He was born in 1927 in the tiny town of Denmark, Iowa, the son of a Congregationalist minister. The family moved from one small rural town to the next when Bob was small; when his father took over the parish in Grinnell, a quiet, attractive town fifty miles east of Des Moines, the family settled there for good. Although the elder Noyces had no particular penchant for technology, they encouraged their sons and listened with interest each night as Bob and his three brothers explained their latest experiments. To pay for his airplane kits, radio parts, chemicals, etc., Bob worked as a baby-sitter and mower of lawns. One of his chief customers was Professor Grant Gale, chairman of the physics department at Grinnell College. Under Gale's tutelage, Noyce fell in love with math and physics. He studied the college texts while in high school, and when he enrolled at Grinnell in 1945 he knew from the first that physics and math would be his major interests. Not his only interests, of course—that was not the way Bob Noyce lived. He was the star diver on Grinnell's swimming team, he sang in choral groups, played oboe in a band, and had a continuing role in a radio soap opera.

Noyce's life story was basically a narrative of one outstanding success after another, but at college he experienced an unforgettable setback. One night in 1948 a group of Grinnell boys decided—it evidently seemed logical at the time—that what their dormitory in the middle of the Iowa plains really needed was a genuine Hawaiian luau, complete with roast whole suckling pig. Tasks were divvied up: Bob and another athletic sort were assigned the job of acquiring the pig—a mission they accomplished by swiping a suckling out of the sty at a nearby farm. The Iowan luau was a great success. But the next morning, when the repentant Noyce confessed his crime, nobody was laughing. In Iowa, theft of livestock is not a humorous matter. Thanks to his father's stature and Grant Gale's intervention, the pork thief was spared a criminal sentence and was kicked out of college for only one semester. He spent the time working at an insurance company. When he reentered Grinnell midway through his senior year he was still able to graduate with his class, earning top grades and making Phi Beta Kappa.

For the most part, the electronics Noyce learned in his physics classes at Grinnell was standard vacuum tube stuff—the Fleming rectifier, the De Forest amplifier, and various improvements to those basic devices. One day, however, Professor Gale astounded the class with news of something totally different. Gale had been a classmate of John Bardeen in the engineering school at the University of Wisconsin, and thus he was able to obtain one of the first transistors and demonstrate it to his students. It was not a lecture the student was to forget. "It hit me like the atom bomb," Noyce recalled forty years later. "It was simply astonishing. Just the whole concept, that you could get amplification without a vacuum. It was one of those ideas that just jolts you out of the rut, gets you thinking in a different way."

Solid-state technology was a whole new world, a whole new universe, for a student of physics. Noyce was bright enough to realize that this had to be the future; there would be no going back to the vacuum tube. As a student of physics at precisely the right moment, Noyce decided to enter this astonishing new world. He applied to MIT and was admitted to a graduate course in physical electronics—and was accepted, of course. He took his Ph.D. there in 1953, sifted through a long list of job offers, and went to work in Philadelphia for Philco, which was just embarking on a large-scale effort to develop and produce improved transistors. The job gave Noyce a chance to practice serious science. He turned out a series of monographs and papers on semiconductor devices. In one paper, Noyce reported on the effects of low-energy gas discharges on a platinum-germanium diode; in another, he demonstrated that the "dc transistor current amplification factor" could be determined using this formula:

$$\alpha = [\text{Sech}\ (W_b/L_b)\ \{\ 1 + 1\ (J_{rq'} + J_{d'})\ /\ J_d\}\ \text{Tanh}\ (W_h/L_b)]^{-1}$$

Late in 1955, Noyce gave a paper before the American Physical Society on "base widening punch-through." That paper caught the eye of the nation's preeminent semiconductor specialist.

On a January day in 1956, Noyce received a telephone call from William B. Shockley. Shockley explained that he was leaving Bell Labs and moving to California to start a new company that would develop high-performance transistors. Would Noyce be interested in interviewing for a job? "It was like picking up the phone and talking to God," Noyce recalled later. "He was absolutely the most important person in semiconductor electronics. Getting that job meant you would definitely be playing in the big leagues." Noyce took a cross-country train to Palo Alto. With characteristic Noycean confidence, he spent the morning of his arrival renting a house near Shockley's lab; that done, he went to the interview to see if he could land the job. He got it, settled his family into the new home, and set to work with Shockley on the development of a high-performance double-diffusion transistor.

The transistor, as Shockley conceived it, is a semiconductor sandwich in which a thin region of P-type silicon (that is, silicon that has been doped with impurities so that it contains extra positive charges) is sandwiched between two regions of N-type silicon (silicon doped with excess negative charges). (There are also less common transistors that employ the opposite structure, with N-type material in the center and P-type on either end.) To get the best performance characteristics from the device, the separate regions of the silicon chip should be clearly defined, and the P-N junction, the point where the different regions meet, should be a sharp, sudden transition from P-type to N-type material. Through the early 1950s transistor makers tried dozens of different techniques to achieve the precise doping and the sharply defined junctions required for reliable transistor action. The process that eventually proved best—the process still used today in semiconductor manufacture—was a Bell Labs discovery called diffusion. Semiconductor diffusion works like a barbecue pit where hickory smoke seeps into the meat and imparts a distinctive flavor. In the diffusion process, a bar of silicon is cooked in a furnace at high heat, and then a gas containing the appropriate doping impurities—boron, for example, or arsenic—is pumped into the furnace. At temperatures of 1000 degrees centigrade or so, some of the impurity atoms in the gas seep into the silicon bar and dope it, either with excess electrons or with positively charged holes. In the same way that a barbecue chef knows just how long to cook the ribs to get the right taste of hickory, solid-state physicists gradually determined the proper time and temperature needed to put precise amounts of impurities at precise points on the silicon block. Thus diffusion provided the first effective means of producing a semiconductor bar with sharply defined regions of

N- and P-type silicon. The process used to make transistors was called double diffusion.

A round wafer of silicon, about the size of a 45 rpm record in diameter and about the thickness of five pages of this book, would be placed in the furnace. Two different kinds of impurities would be diffused onto the wafer in separate steps, leaving a three-layer cake—N-type on the bottom, P-type in the middle, N-type on top. The wafer could then be cut, like a cake, into dozens or scores of tiny three-layer pieces—each one an N-P-N transistor.

Double diffusion made possible, for the first time, the mass production of precise, high-performance transistors. The technique promised to be highly profitable for any organization that could master its technical intricacies. Shockley therefore quit Bell Labs and, with financial backing from Arnold Beckman, president of a prestigious maker of scientific instruments, started a company to produce double-diffusion transistors. The inventor recruited the best young minds he could find, including Noyce; Gordon Moore, a physical chemist from Johns Hopkins; and Jean Hoerni, a Swiss-born physicist whose strength was in theory. Already thinking about human intelligence, Shockley made each of his recruits take a battery of psychological tests. The results described Noyce as an introvert, a conclusion so ludicrous that it should have told Shockley something about the value of such tests. Early in 1956, Shockley Semiconductor Laboratories opened for business in the sunny valley south of Palo Alto. It was the first electronics firm in what was to become Silicon Valley.

In Robert Noyce's office there hung a black-and-white photo that showed a jovial crew of young scientists offering a champagne toast to the smiling William Shockley. The picture was taken on November 1, 1956, a few hours after the news of Shockley's Nobel Prize had reached Palo Alto. By the time that happy picture was taken, however, Shockley Semiconductor Laboratories was a chaotic and thoroughly unhappy place. For all his technical expertise, Shockley had proven to be an inexpert manager. He was continually shifting his researchers from one job to another; he couldn't seem to make up his mind what, if anything, the company was trying to produce.

"There was a group that worked for Shockley that was pretty unhappy," Noyce recalled many years later. "And that group went to Beckman and said, hey, this isn't working. . . . About that time, Shockley got his Nobel Prize. And Beckman was sort of between the devil and the deep blue sea. He couldn't fire Shockley, who had just gotten this great international honor, but he had to change the management or else everyone else would leave." In the end, Beckman stuck with Shockley—and paid a huge price.

Confused and frustrated, eight of the young scientists, including Noyce, Moore, and Hoerni, decided to look for another place to work. That first group—Shockley called them "the traitorous eight"—turned out to be pioneers, for they established a pattern that has been followed time and again in Silicon Valley ever since. They decided to offer themselves as a team to whichever employer made the best offer. Word of this unusual proposal reached an investment banker in New York, who offered a counterproposal: Instead of working for somebody else, the eight scientists should start their own firm. The banker knew of an investor who would provide the backing—the Fairchild Camera and Instrument Corporation, which had been looking hard for an entrée to the transistor business. A deal was struck. Each of the eight young scientists put up $500 in earnest money, the corporate angel put up all the rest, and early in 1957 the Fairchild Semiconductor Corporation opened for business, a mile or so down the road from Shockley's operation.

Noyce recalled that the group had some slight qualms about running their own business, but these doubts were easily overcome by "the realization, for the first time, that you had a chance at making more money than you ever dreamed of." The dream, as it happened, came true. Even by high-tech standards, that $500 turned out to be a spectacular investment. In 1968 the founders sold their share of Fairchild Semiconductor back to the parent company; Noyce's proceeds—the return on his initial $500 investment—came to $250,000. Noyce and his friend Gordon Moore had by then found another financial backer and started a new firm, Intel Corporation (the name is a play on both *Intelli*gence and *Inte*grated *Ele*ctronics). Intel started out making chips for computer memories, a business that took off like a rocket. Intel's shares were traded publicly for the first time in 1971—on the same day, coincidentally, that Playboy Enterprises went public. On that first day, stock in the two firms was about equally priced; a year later, Intel's shares were worth more than twice as much as Playboy's. "Wall Street has spoken," an investment analyst observed. "It's memories over mammaries." Today, Intel is a multibillion-dollar company, and anybody who held on to the founding group's stake in the company is a billionaire several times over.

The men who started Fairchild Semiconductor in 1957 were determined to make the double-diffusion transistor, but they were also on the outlook for any other product that could turn a profit. Noyce, as director of research and development, was the man responsible for spotting important technical problems that Fairchild might profitably solve. And it didn't take a genius, at this point in technological history, to identify the most important problem by far facing the electronics industry: the tyranny of numbers. Off and on, all through 1957 and 1958, Noyce thought about the interconnections problem. In retrospect, he said later, the monolithic idea should have come to him much earlier. "Here we were in a factory that was making all these transistors in a perfect array on a single wafer," Noyce said, "and then we cut them apart into tiny pieces and had to hire thousands of women with tweezers to pick them up and try to wire them together. It

just seemed so stupid. It's expensive, it's unreliable, it clearly limits the complexity of the circuits you can build. It was an acute
problem. The answer, of course, was, don't cut them apart in the
first place. But nobody realized that then." Instead, Noyce was
stuck in a rut. He worked on standard ideas for making circuit
components in smaller sizes and uniform shapes. He came up
with nothing worthwhile.

Unable to get out of the rut, Noyce turned his attention to another technical dilemma. Double-diffusion transistors—the tiny
three-layer chips of N-P-N silicon—were highly susceptible to
contamination. A piece of dust, a stray electric charge, a minute
whiff of contaminating gas would break down the P-N junctions
and impair transistor action. One day in 1958, Jean Hoerni came
to Noyce with a theoretical solution: He would place a layer of silicon oxide on top of the N-P-N chip, like icing atop the three-
layer cake. The oxide would hold fast to the silicon and protect it
from contaminants. "It's building a transistor inside a cocoon of
silicon dioxide," Noyce explained, "so that it never gets contami-
nated. It's like setting up your jungle operating room. You put the
patient inside a plastic bag and you operate inside of that, and you
don't have all the flies of the jungle sitting on the wound."

The people at Fairchild recognized Hoerni's idea—it was
called the planar process because it left a flat plane of oxide atop
the silicon—as an important advance in transistor technology.
Noyce quickly called in the firm's patent lawyer, John Ralls, to put
together a patent application. Ralls, sensing that this planar idea
might have other applications in electronics, wanted to write the
application in the broadest language. He told Noyce that
Fairchild should make the application as expansive as it possibly
could. Every time they talked about the planar patent, Ralls
would pose a challenge: "What else can you do with this idea?"

Looking back years later, Noyce could see clearly that it was the
lawyer's question that pushed him out of his mental rut and pro-

voked the leap of insight that became the monolithic idea. What else? What else could you do? In the first weeks of 1959, Noyce was thinking hard about that question, scratching pictures in his notebook, talking things over hour after hour with his sedate, cautious friend Gordon Moore.

As Noyce looked over Hoerni's planar idea, he realized that it had another useful property. It was quite difficult in those days to make precise electrical connections to the separate regions of an N-P-N transistor, because wires were relatively large compared to the tiny regions of the chip. Hoerni's oxide icing, spread atop the three-layer silicon cake, helped solve this problem. Connecting wires could be poked down through the icing like candles on a cake, and they could be inserted at the exact spot on the chip where the connection was needed. The oxide would keep them firmly in place. "Remember, what I'm trying to do is make this [transistor] extremely small," Noyce explained. "Well, I can't attach a wire to that, because it's too small. But now, with the planar coating, I can attach a big old wire—big old wire, this is, you know, a quarter of a human hair—running it on top of the oxide."

And that realization led to a new idea, something that was even better: wires wouldn't be needed at all. Noyce now saw that tiny lines of copper or some other metal could be printed on top of the oxide layer rather than poked down through. The advantage of this approach was that printing is a faster industrial process than aligning and inserting tiny wires. With the "wires" printed on top of the oxide coating, all the transistor's interconnections, all the wires, could be made in one fell swoop in a single manufacturing process.

But wait a minute, Noyce thought. We can carry this even further. If you could connect the separate regions of a single transistor with these printed metal lines, then you could put two separate transistors on a single piece of silicon and connect them with the printed lines. And why stop with transistors? If you could put two transistors on a single chip of silicon, couldn't you build some other circuit components on the same chip? How about a resistor built into the same chip? How about a capacitor in there? Couldn't you, in fact, build a complete circuit, an integrated circuit, all on a single chip of silicon? Wouldn't that overcome the tyranny of numbers?

"I don't remember any time when a light bulb went off and the whole thing was there," Noyce said. "It was more like, every day, you would say, Well, if I could do this, then maybe I could do that, and that would let me do this, and eventually you had the concept."

One day, Noyce walked into Moore's office and showed him, on the blackboard, that two transistors in a single silicon block could be connected by printed copper lines on the oxide layer. A few days later, he was back at the blackboard, showing Moore how he could use a channel of undoped silicon in the same block as a resistor. A few days later, he was drawing a silicon capacitor on the blackboard. It was all completely new, but Moore raised no serious objections.

Noyce's intellectual journey to the monolithic idea began at a different starting point from Jack Kilby's, but reached the same destination. Kilby had first hit on the (crazy) idea of building all the circuit elements in a single semiconductor block; as an addendum to that notion, he realized that the various elements could be connected by "wires" printed onto the same block. Noyce, on the other hand, arrived first at the idea of printing the wires on the semiconductor chip, and went from that level to the idea of putting all the circuit elements on a chip. Both routes led to the monolithic integrated circuit.

On January 23, 1959, "all the bits and pieces came together," and Noyce filled four pages of his lab notebook with a remarkably complete description of an integrated circuit. "In many applications," he wrote, "it would be desirable to make multiple devices on a single piece of silicon, in order to be able to make interconnections between devices as part of the manufacturing process, and thus reduce size, weight, etc. as well as cost per active element." Noyce went on to explain how resistors and capacitors could be fabricated on a silicon chip, and how the whole monolithic circuit could be connected by metal contacts printed right onto the chip. He also set forth a rough sketch of a computer circuit—a basic "adder" circuit that would add two numbers—realized in integrated form. Six months after Jack Kilby had arrived at the monolithic idea, Bob Noyce sailed into the same port. Kilby's journey had been slightly quicker, but the use of the planar process made the Noyce route somewhat more viable.

News travels quickly in the electronics industry. By the spring of 1959, rumors about a major new development at Texas Instruments had reached the people at Fairchild. Nobody knew exactly what TI had done, but it was not impossible to guess which problem this breakthrough was designed to solve. Noyce again called in John Ralls and asked the lawyer to prepare a patent application for a new idea—"a unitary circuit structure . . . to facilitate the inclusion of numerous semiconductor devices within a single body of material." This time, Ralls decided to write a detailed, precise patent application, a document that could serve as a shield to protect Fairchild against any possible legal action by Texas Instruments. This strategic decision would become the decisive factor in a bitter ten-year legal battle fought all the way to the United States Supreme Court.

KILBY V. NOYCE

The terrifying rumor that raced through the semiconductor lab at Texas Instruments on the morning of January 28, 1959, turned out, eventually, to be wrong on almost every count. But like many false alarms, it had the salutary effect of scaring people into action. More than four months had passed since Jack Kilby had successfully demonstrated his prototype integrated circuit, but since then further development of the concept had been almost nil. Kilby's superiors had been hoping to introduce this fantastic new product in March, but as of the end of January, there really wasn't any product. The only integrated circuits in existence were the crude models Kilby had built by hand for his demonstration; nobody had figured out yet how to turn out a production version. Even the lawyers were behind schedule; they had failed to take the most basic steps to protect Texas Instruments' right to the new invention.

That's why the rumor was so frightening. At a technical meeting, somebody from TI thought he heard somebody else saying that he had been told that another company had come up with an integrated semiconductor circuit that would overcome the tyranny of numbers. There was, in fact, a germ of truth in this report; just five days earlier, Bob Noyce, in his office at Fairchild, had scratched his first sketchy concept of the monolithic idea in his notebook. But the rumor that reached Dallas had nothing to do with Noyce. The word at TI was that somebody at RCA had come up with an integrated circuit, and that—even worse—RCA was soon going to file for a patent. When this unsettling news reached Samuel M. Mims, TI's senior staff lawyer, he didn't hesitate a second before putting through an emergency call to Mo Mosher.

Ellsworth H. Mosher, name partner in the Washington, D.C., patent law firm of Stevens Davis Miller & Mosher (cable address: INVENTION) and an elder statesman of the patent bar, knew immediately that he had to act fast. He dispatched a junior lawyer to Dallas and told Mims to sit down with Kilby and find out precisely what the inventor thought his monolithic idea would be good for. It normally took two or three months to put together all the paperwork, prose, and pictures required for a patent application; in this case, though, Mosher promised to deliver a completed application to the Patent Office within a week.

Mosher's advice—that Texas Instruments had better apply for a patent, and fast—was not quite so obvious as it might appear. One of the most important rules that patent lawyers try to get across to their clients is that, in some cases, it is better not to apply for a patent at all. For an inventor a patent is a sort of Faustian bargain.

The patent expressly guarantees the inventor "the right to exclude others from making, using, or selling" the idea for the twenty-year life of the patent. The patent holder can, if he chooses, issue licenses to others to make, use, or sell the idea. The license fees can bring in large sums of money. If anybody tries to market the patented product without obtaining a license, the inventor can go into federal court to get an injunction and money damages. Not a bad deal at all for the inventor. In exchange for those benefits, though, the patent holder has to reveal all the secrets of his success. The patent law says that an inventor must provide "a written description of the invention, and of the manner and process of making and using it, in . . . full, clear, concise and exact terms." The inventor and his company might have expended a dozen years and a hundred million dollars perfecting the idea; once a patent is granted, anybody in the world can acquire the plans— full, clear, concise, and exact—from the Patent Office for $3.

If, for example, John S. Pemberton had applied for a patent for the formula he whipped up in his backyard in Atlanta one day in the mid-1880s, the product that he invented—a soft drink that he named Coca-Cola—would have entered the public domain in 1903, when the patent expired. Anybody in the world would have been free from that day forward to brew and sell the drink without paying a penny to the Coca-Cola Company. But Pemberton kept his formula unpatented, and thus secret. Even without a patent, Coca-Cola has been able to defend its formula under a body of law known as trade secret protection, which makes it illegal to copy deliberately somebody else's commercial idea.

From the inventor's viewpoint, the flaw with the trade secret laws is that they apply only to purposeful stealing of an idea. They do not prevent anybody from marketing a product that he has invented on his own, even if an earlier inventor has been selling the same product for years. Lacking a patent, Coca-Cola would have no recourse against a company selling exactly the same drink if the second firm could prove in court that its chemists had been messing around with sugar, flavorings, and cola nuts and just happened to hit on the precise formula that Coca-Cola uses. The holder of a patent, in contrast, can go to court to stop any competitor from selling the same product, even if the competitor developed the product completely on his own. The strategic decision facing every inventor, then, is whether he wants twenty years of the stronger protection provided by a patent, or permanent protection under the trade secret laws against only those who deliberately steal the idea.

The choice has to be made, because an inventor can take either patent protection or trade secret protection, but not both. This principle was established once and for all in the landmark case of *Kellogg* v. *Nabisco*—the great shredded wheat decision. The familiar shredded wheat biscuit was invented in 1895 by a Colorado baker named Henry Perky, who promptly—and foolishly, as it turned out—took out a patent (No. 548,086) on his new breakfast cereal. The National Biscuit Company subsequently bought the rights to Perky's invention. Despite extensive advertising, shredded wheat sales never took off until the 1920s—well after the patent had expired. Somebody in Battle Creek realized that the formula was free for the taking, and grocers everywhere began carrying a new product—Kellogg's Shredded Wheat. Nabisco went to court, claiming, under the trade secret laws, that Kellogg had deliberately copied its product. The shredded wheat litigation wound its way for years through the legal system, and in 1938 it finally reached the United States Supreme Court. In a decision by Louis Brandeis, the Court sided with Kellogg all the way. Once the patent had been issued, Brandeis wrote, Nabisco lost its right to claim that

> shredded wheat was its private secret. The basic goal of the patent system, after all, was to encourage public disclosure of technological advances.

This goal was so important to the founding fathers that the Patent Office (the name comes from the Latin verb *patere*, "to open") was one of the earliest federal agencies created by the First Congress. Among the small group of men who constituted the U.S. government then, jobs were assigned largely on the basis of personal interest. The secretary of state, Thomas Jefferson, was an inventor, so Congress gave the Patent Office to the State Department. In the evening, after a long day of diplomacy, Jefferson would review patent applications. Among the patents he personally granted was one to Eli Whitney in 1794 for the cotton gin.

Inevitably, bureaucracy reared its head. In 1804 the Patent Office hired an employee of its own, and that proved to be the first step along a steep slope; today the Patent and Trademark Office (now a wing of the Commerce Department) has about 7,000 employees scattered around several huge buildings in a suburban mall in Arlington, Virginia.

The government receives about 270,000 patent applications each year and grants about 170,000 patents for inventions, plus more for plants and designs. The office is, among other things, one of the world's busiest publishing houses; it keeps every one of the 6.2 million patents issued since Jefferson's time in print.

Like most traditional institutions, the Patent Office has developed parlance and procedures all its own. There are tens of thousands of pages of statutes, regulations, guidelines, and legal opinions governing the issuance of patents. Among much else, the regulations go on for fifteen long paragraphs describing the kind of paper and ink an applicant should use for drawings of his invention: "The sheets may be provided with two ¼-inch (6.4 mm) diameter holes having their centerlines spaced ¹¹⁄₁₆ inch (17.5 mm) below the top edge and 2¾ inches (7.0 cm) apart, said holes being equally spaced from the respective side edges."

At the core, though, the basic rules governing what kind of inventions will be granted a patent are straightforward. The invention has to be "new." It has to be "useful," a term the courts have interpreted to mean that the gadget has to work. The Patent Office receives a few applications each year from people who have invented perpetual motion machines; it rejects them all on the ground that they can't do what they're supposed to and so aren't useful. The inventions also have to be desirable; the government refuses to grant patents for nuclear weapons, no matter how new or how "useful" they might be.

When Mo Mosher heard about the important new invention at Texas Instruments, he knew immediately that TI would need a patent. Unlike soft drinks and breakfast cereals, electronic gear rarely has a market life longer than the term of a patent, so it is almost always a wise move for an electronics inventor to seek a patent. Mosher knew, too, that none of the statutory requirements that govern patentability would pose a problem. Kilby's monolithic circuit was clearly something completely new, and since it promised a solution to the most important problem facing the industry, it was eminently useful and desirable. And relying on trade secret protection in this case was unlikely to be a smart bet; so many other people around the world were looking for a solution to the tyranny of numbers that somebody was likely to come up with an integrated circuit without any access to TI's internal secrets. A patent meant TI could own rights to the chip even if somebody else independently hit on the monolithic idea.

Before he could start writing Kilby's application, though, Mosher had to resolve a fundamental tactical question. Anyone who applies for a patent has to decide whether he needs it for offensive or for defensive purposes—whether, to use lawyers' favorite metaphor, he wants his patent to be a sword or a shield. The decision usually turns on the novelty of the invention. If somebody has a genuinely revolutionary idea, a breakthrough that his competitors are almost sure to copy, his lawyers will write a patent application they can use as a sword; they will describe the invention in such broad and encompassing terms that they can take it into court for an injunction against any competitor who tries to sell a product that is even remotely related. In contrast, an inventor whose idea is basically an extension of or an improvement on an earlier idea needs a patent application that will work as a shield—a defense against legal action by the sword wielders. Such a defensive patent is usually written in much narrower terms, emphasizing a specific improvement or a particular application of the idea that is not covered clearly in earlier patents.

Probably the most famous sword in the history of the patent system was the sweeping application filed on February 14, 1876, by a teacher and part-time inventor named Alexander Graham Bell. That first telephone patent (No. 174,465) was so broad and inclusive that it became the cornerstone—after Bell and his partners had fought some 600 lawsuits against scores of competitors—of the largest corporate family in the world. In the nature of things, though, few inventions are so completely new that they don't build on something from the past. The majority of patent applications, therefore, are written as shields—as improvements on some earlier invention. Some of the most important patents in American history fall into this category, including No. 586,193, "New and Useful Improvements in Transmitting Electrical Impulses," granted to Guglielmo Marconi in 1898; No. 621,195, "Improvements in and Relating to Navigable Balloons," granted to Ferdinand Zeppelin in 1899; No. 686,046, "New and Useful Improvements in Motor Carriages," granted to Henry Ford in 1901; and No. 821,393, "New and Useful Improvements in Flying Machines," granted to Orville and Wilbur Wright in 1906.

The integrated circuit, however, was so new, and so potentially lucrative, that Mosher, Mims, and Kilby decided to shoot the works—to write a far-reaching patent application that Texas Instruments could wield as an offensive weapon against anyone else who tried to make, use, or sell the device. Accordingly, Mosher and his partners set out to write a document that would leave no doubt about the revolutionary quality of Kilby's idea. They laid it on thick. Every aspect of the integrated circuit was described as "novel" or "unique," or both. "Radically departing from the teachings of the art," the application said, ". . . the present invention has resulted from a new and totally different concept for miniaturization." As a result "the ultimate in circuit miniaturization is attained." And for good measure: ". . . the invention . . . represents a remarkable improvement over the prior art."

The document went on to describe two specific circuits that could be fabricated on a semiconductor chip. But Mosher emphasized that Kilby's invention could be used to build an integrated version of any circuit: "There is no limit upon the complexity or configuration of circuits that can be made in this manner." To put a point on his sword, Mosher added a thinly veiled warning to anybody who might try to circumvent the patent by making small changes: "Such changes and modifications," the application said, "are deemed to fall within the purview of this invention."

For all this rhetorical bravado, the authors of Kilby's patent application still had to contend with a fairly sticky problem. Since the patent system is designed to let the world know the secrets behind technological advances (and thus, in theory, stimulate new advances), the law requires an inventor to explain in his patent application how the device is made. "The specification," the statute says, ". . . shall set forth the best mode contemplated by the inventor for carrying out his invention." Further, the law requires an inventor to provide drawings showing precisely what his new device looks like. This was a problem because nobody at Texas Instruments, not even Jack Kilby, had figured out the best mode for carrying out the invention. And nobody knew what a production-model integrated circuit would look like.

In a way, Kilby's invention was almost too ingenious, too violent a break with the way things had always been done before. By the late 1950s, engineers had been building electric circuits for almost 100 years, and they had always done it the same way: taking individual resistors, capacitors, and other components and wiring them together. The invention of the transistor had led to dramatic changes in the size, cost, reliability, and efficiency of circuits, but it had not changed the basic structure of discrete components wired together. That was one reason the tyranny of numbers had proven such a baffling problem; to the engineers, a large number of discrete parts were what circuits were all about.

The monolithic idea—the idea that an entire circuit could be a single part, with resistors, capacitors, diodes, and transistors built right in—involved two fundamentally new concepts, which Kilby later defined as *integration* and *interconnection*. Integration was the central idea: the realization that all the parts of a circuit could be made from the same material—silicon—and thus a whole circuit could be integrated in a single silicon chip. Interconnection was the recognition that, once all the parts of a circuit were made on a single chip, the connections between the different parts— the wires—could be printed onto the chip as part of the production process. That way, the enormous cost and the inherent unreliability of hand-wiring huge numbers of components to one another would be eliminated.

In the first chips he built, during late summer and fall of 1958, Kilby successfully integrated all the circuit components into a silicon chip. But he didn't have time to work out the interconnections within the chip. Racing to demonstrate the new idea to his bosses before they lost interest, he had settled for a jury-rigged device in which the separate components on the chip were connected (by hand) with tiny gold wires, giving that first integrated circuit the appearance of an intricate cobweb spun by a golden spider. This crude handmade circuit was a mess—just a sloppy piece of lab equipment, like the dirty glass tube with wires sticking out that was Edison's first light bulb. The world's first integrated circuit had been adequate for its purpose—proving that a circuit on a chip would really work—but the job of attaching the wires was much too painstaking and time-consuming for any large-scale production. Before the integrated circuit could become a practical solution to the numbers barrier, somebody would have to deal with the question of interconnection.

By the beginning of 1959, Kilby had somewhat improved his

techniques for constructing individual components within a semiconductor chip. With several TI colleagues, he had worked out some reliable designs for different circuits on a chip. But the interconnections business was still up in the air. Kilby had a vague idea, an idea that was similar to the planar process that Noyce's colleague Jean Hoerni had developed at Fairchild. He thought he could put a layer of silicon oxide on top of the chip, like icing on a cake, and then print fine lines of a conducting metal, such as gold, atop the icing to connect various parts of the circuit. (This was, in fact, the method eventually developed for chip production.) But at the end of January, the only chips in existence still used the handmade cobwebs of wire.

Here was a dilemma: Worried that RCA was going to apply for a patent any minute, the TI people were anxious to file first. But a patent application required a picture. To draw a picture, Mosher's artist needed a model. But the only model available was Kilby's crude demonstration chip, with its network of gold wire. After a week or so of deliberation, it was decided to put speed first; the application was filed with a picture of the hand-wired chip:

Even now, more than forty years later, engineers tend to smirk when they see this drawing—it is known in the electronics business as the "flying wire picture"—because it is so far removed from what an integrated circuit is supposed to be.

Jack Kilby, of course, was one of the engineers who knew that the flying wire drawing was fundamentally wrong. Accordingly, the lawyers threw in some language designed to fudge the issue.

"Although the invention has been shown and described in terms of specific embodiments," they wrote, "it will be evident that changes and modifications are possible which do not in fact depart from the inventive concepts taught herein."

That helped a little, but Kilby still wasn't satisfied; he had to put in something to show that his revolutionary new circuit would not have to be wired by hand. At the last minute, accordingly, Kilby and Mims added one more paragraph to the application, setting forth the icing-on-the-cake approach to interconnection. "Instead of using the gold wires in making electrical connections," it said, "connections may be provided in other ways. For example . . . silicon oxide may be evaporated onto the semiconductor circuit wafer. . . . Material such as gold may then be laid down on the [oxide] to make the necessary electrical connections." With this final addition, the application—four pages of pictures and five of text—was delivered to the Patent Office on February 6, 1959.

There, Kilby's application for a patent on his "Miniaturized Electronic Circuits" was assigned to an examiner who specialized in inventions involving electronic circuits and devices. The Patent Office employs about 2,500 examiners (they are generally lawyers with a background in some technical field) whose job is simple to describe—they have to determine whether an invention is legally entitled to a U.S. patent—but often extremely complicated to perform. The job of patent examiner tends to attract people with a curious, probing intelligence; among the trade's most illustrious alumni was a hard-working physicist who held the position of "Technical Expert, Third Class" at the Swiss Patent Office in Bern at the turn of the twentieth century—a fellow named Albert Einstein.

The examiner has to determine whether an idea is "new"—a question that requires searching dozens, hundreds, or thousands of earlier patents, reports, and monographs. He has to decide whether it is "useful," a determination that demands extensive research in the technical literature. And he has to determine whether the inventor's application clearly sets forth how the new gadget is to be built. On the average, each examiner is assigned between 75 and 100 patent applications every year. As a result, it regularly takes months, and sometimes years, for an examiner to deliver his initial opinion on an application. If the examiner finds anything amiss—and he almost always does—the Patent Office sends a letter to the inventor explaining the reservations. The inventor gets up to six months to prepare a reply or amend the application; then the examiner may take months more to review the reply. It is not uncommon, particularly for a complex invention in a specialized field, for an inventor to wait years for final action on a patent application.

So it was for Kilby. The process dragged on, with the examiner raising questions and Texas Instruments doing its best to answer them. The first important development came some twenty-six months after the application had been filed. On April 26, 1961, Kilby received a telephone call from the lawyers in Washington, informing him that the first patent ever for an integrated circuit had been granted.

But it had not been granted to Jack Kilby.

Like their counterparts at Texas Instruments, the people at Fairchild Semiconductor took their sweet time at first about developing the monolithic idea into a practical integrated circuit. It was nearly two months after Bob Noyce had set down the basic concept in his notebook before Fairchild started working on the idea; another four months passed before Noyce got around to filing for a patent.

The reason, Noyce explained later, was that, after twenty months of preliminary planning and organizing, Fairchild at the start of 1959 was just beginning to sell its first important product—

the double-diffuse transistor. "We were still a brand-new company," Noyce recalled. "We were worried about basic survival. That meant getting transistors out the door. The integrated circuit seemed interesting, it was something that might make you some money somewhere down the road, but that was not a period when you had a lot of time for it." During the winter of 1959, the monolithic idea was literally put on a shelf; except for his friend and sounding board, Gordon Moore, Noyce didn't even mention it to any of his colleagues.

Like their counterparts at Texas Instruments, the people at Fairchild were eventually prodded into action by a rumor—although in this case, the rumor was true. Sometime in late February or early March, word arrived in Silicon Valley that Texas Instruments was about to announce a wholly new kind of circuit that would do away with discrete electric components by integrating all the parts of a circuit into a single silicon chip. For any company that made its money selling discrete components, like Fairchild's transistors, this was disconcerting news. Their basic product was suddenly faced with obsolescence. Somebody at Fairchild called a meeting to discuss the development. At this session Noyce explained for the first time that he, too, had worked out the basic concept of an integrated circuit. The group decided immediately to get hopping on the new product. To any engineer looking at plans for a possible new product, it was obvious that one of the first steps would be to file for a patent.

Noyce and his patent lawyer, John Ralls, did not have access to Kilby's patent application; the Patent Office treats pending applications as classified material, and there are no leaks. But by mid-March, when TI publicly announced its new "Solid Circuit," they knew that the Dallas firm must have filed already. What Fairchild needed, then, was a legal shield—a patent that would differentiate Noyce's version of the idea from the Kilby invention and thus permit Fairchild to enter the integrated circuit market without fear of legal action by TI.

In preparing his application, Noyce had one significant advantage: Kilby had been forced to file for a patent before he had worked out the problem of interconnections within the chip; Noyce's formulation of the idea covered both integration and interconnection. This was a result of the different routes the two inventors had taken to arrive at the integrated circuit. Kilby had first hit upon the concept of integration—of building all the parts of a circuit in a monolithic chip of silicon—and had moved from there to consideration of interconnections. Noyce, in contrast, had first recognized the possibility of printing connecting strips of metal on a chip—something made possible by Jean Hoerni's invention of the planar process—and the notion of interconnection had led him to the idea of integration. By the spring of 1959, Fairchild was busily engaged in working out the details of Hoerni's planar process, and thus Noyce had no difficulty providing a description and a drawing of a chip with interconnections built right in. Accordingly, Noyce and Ralls titled their application "Semiconductor Device-and-Lead Structure" ("lead" is the electrician's term for a connecting wire in a circuit), and they strongly emphasized the interconnections aspect of the Noyce circuit. The application listed three "principal objects" of the invention. The first one was interconnections: "to provide improved device-and-lead structures for making electrical connections to the various semiconductor regions."

Noyce's application went on to describe a "unitary circuit structure" that would permit integrating "more than one circuit device, into a single body of semiconductor."

"According to prior practice," it continued, "electrical connection . . . had to be made by fastening wires directly to the [components]. . . . By means of the present invention, the leads can be deposited at the same time and in the same manner as the [components] themselves."

To the four-page written description of Noyce's idea, Ralls added three pages of pictures of typical circuits that could be integrated onto a chip. There were no flying wires. There were no wires at all. The drawings show a structure that is essentially the same as the integrated circuits being produced today.

It was midsummer before Ralls and Noyce were satisfied that everything was in order; the patent application was finally filed on July 30, 1959. It was assigned, evidently, to an examiner who was not aware of the earlier application from Texas Instruments, and it moved ahead at what is, for the Patent Office, lightning speed. Twenty-one months after it was filed, Noyce's application was granted: U.S. Patent No. 2,981,877. By formal decree of the United States of America, Robert N. Noyce—the second person to come up with the idea—had been officially declared the inventor of the integrated circuit.

The award of an integrated circuit patent to Noyce evoked consternation, but not outright panic, at Texas Instruments. Kilby and his lawyers, after all, were veterans of the patent game; they knew that some applications move through the Patent Office faster than others, and that it is not particularly unusual for the second version of an invention to be the first patented. This happens so often, in fact, that the government has a special procedure— called an interference proceeding—and a special board—the Board of Patent Interferences—to consider the claims of inventors who find themselves in Kilby's position. The basic rule governing an interference is that priority prevails—that is, whichever inventor can prove to have had the idea first gets the patent.

Mosher filed the necessary papers, and in May 1962 both Noyce and Kilby received a copy of Commerce Department

Form POL-102, declaring that the Board of Patent Interferences had convened Interference No. 92,842, "*Kilby* v. *Noyce*," to determine who had really been the first to invent the microchip. The board enclosed a short form asking each man to list the earliest date on which he could prove he had had the idea. Since both Kilby and Noyce had maintained lab notebooks precisely for this purpose, both were able to provide an exact answer: July 1958 for Kilby, January 1959 for Noyce. With these preliminaries concluded—they consumed about ten months—the stage was set for a final determination.

Actually, it was not quite set. Before the central legal battle could get under way, the lawyers fought out a series of preliminary skirmishes:

"Motion to Dissolve Under Rule 232 (a) (2)"
"Opposition to Motion to Dissolve Under Rule 232 (a) (2)"
"Motion to Dissolve Under Rule 232 (a) (3)"
"Opposition to Motion to Dissolve Under Rule 232 (a) (3)"
"Request to File Affidavits"
"Opposition to Motion to File Affidavits"

While this was going on, the Patent Office concluded that Jack Kilby's initial application was satisfactory after all. In June 1964, Kilby was granted a patent—No. 3,138,743—for the integrated circuit. This made the interference proceeding even more crucial, and the lawyers went back to work:

"Motion for Extension of Time"
"Opposition to Motion for Extension of Time"

Each of these preliminary disputes took a few months to resolve. And so it was not until July 28, 1964—more than two years after the interference was begun—that the inventors and their lawyers gathered in Mosher's Washington office to hear the first piece of evidence in the case.

The session was a brief one, directed strictly to prove that Kilby had made the invention first. Kilby explained that he had gotten the idea in July 1958, and one of Mosher's associates related how he had filed Kilby's patent application in February 1959. There was a short cross-examination, and everybody went home.

Three months later, everyone met again in a lab in Palo Alto to hear Fairchild's response. There was not a great deal the Fairchild people could say on the basic question, priority of invention. On that point, Kilby was a clear winner. Nonetheless, before the session ended, Noyce's lawyers had dropped a bombshell.

During the long months of preliminary backing and filling, Fairchild's trusted patent attorney, John Ralls, had died. His place was taken by Roger Borovoy, a junior member of Ralls's firm who had caught Noyce's eye and eventually won his confidence. A lot of young lawyers might have found Borovoy's position somewhat daunting. In his first big case he was litigating against Mosher, a titan of the bar, and the facts on the crucial issue were all on Mosher's side. As Borovoy saw it, however, he was sitting pretty. "Here I was," he recalled later, clearly savoring the memory, "a punk kid, defending the most important electronics patent in twenty-five years and Mo Mosher opposing me. Fantastic, right?"

Since he had no reasonable defense to Kilby's claim of priority, Borovoy decided to go on the offensive. He pored over Kilby's patent application. What could he attack? Where was the weak spot? And of course, he found one: the flying wire picture. By 1964, when Borovoy took over the case, the industry had largely determined what an integrated circuit would look like; it didn't look anything like the drawing in Kilby's patent application. Focusing on that drawing, Borovoy drew up an offensive strategy. If he could discredit Kilby's application because of its weakness on interconnections, Noyce would be left with the only valid patent for an integrated circuit. Of course, Kilby's application also contained that last-minute paragraph explaining how, in place of the flying wires, "conducting material such as gold may then be laid down on the insulating [oxide] to make . . . connections." To win the case, Borovoy knew, he would have to find something wrong with that paragraph.

Accordingly, when the litigants assembled at Palo Alto to hear the Fairchild testimony, Borovoy brought forward an expert witness—an electrical engineering professor at Stanford—who declared that no one could build an integrated circuit by following the instructions in Kilby's application. The hand-wired circuit in the picture was obviously wrong. For that matter, the business about laying down gold on an oxide was faulty, too. You can lay down gold on oxide, the expert testified, but "it will not stick."

Under Borovoy's gentle prodding, the expert contrasted Kilby's language—"laid down on"—with the wording of Noyce's patent, which said the connection material had to be "adherent to" the oxide layer. "Laid down on" had no clear meaning, the expert said. "Adherent to," in contrast, was a precise technical term. On that fine distinction Fairchild would have to rest its case.

A month later, when inventors and lawyers gathered again at a lab in Dallas to hear Texas Instruments' rebuttal, Mosher produced an expert of his own. This was an engineer from Kilby's alma mater, the University of Illinois, and he thoroughly disagreed with the Stanford man. Gold *will* stick to an oxide layer, he said, so there was no practical difference between "laid down on" and "adherent to." With this testimony, both sides had had their say. All the expert testimony had consumed six more months, but the stage was set for final resolution.

Actually, it was not quite set. First, there were a few more procedural battles to be fought:

"Request for Sur-Rebuttal Testimony"
"Opposition to Request for Sur-Rebuttal Testimony and Con-
ditional Request for Sur-Sur-Rebuttal Testimony"
"Reply to Request for Suspension of Action on Request for
Leave to Take Sur-Rebuttal Testimony and on Conditional
Motion to Take Sur-Sur-Rebuttal Testimony"

Next, the lawyers had to argue the case before the Board of
Patent Interferences. In oral argument and in their written briefs,
both sides gave most of their attention to the interconnections
question. Borovoy's brief included an oversize copy of the flying
wire picture. "Note that this drawing shows no oxide layer and no
gold wires 'laid down on' any such layer," the brief said. "In fact,
it is readily apparent that the gold wires are anything but 'laid
down.' "

Six months later, however, on February 24, 1967, when the
board issued its opinion, it brushed all that aside. After reviewing
the experts' disagreement over "laid down on" and "adherent to,"
the board observed that "we are not particularly impressed with
that testimony." As the board saw it, Kilby's patent application,
while not perfect, was clear enough. That left only the question of
which inventor was first: "Since Noyce took no testimony to establish any date prior . . . Kilby must prevail." Eight years after he
had filed his patent application, Jack Kilby had been adjudicated
the inventor of the integrated circuit. The stage was now set for
Texas Instruments to wield its sword against the rest of the electronics industry.

Actually, it was not quite set. Any American who is unhappy
with a federal agency's decision has the right to appeal, and
Fairchild exercised that right. A year was devoted to the prepara-
tion of briefs and the filing of motions, and in the fall of 1968,
Mosher and Borovoy appeared before the Court of Customs and
Patent Appeals to argue all the issues once again. Another year

passed. On November 6, 1969, the court issued its opinion. This time, the decision dealt exclusively with the difference between "laid down on" and "adherent to." The judges had found Roger Borovoy's argument appetizing—and swallowed it whole. "Kilby has not demonstrated," the opinion said, "that the term 'laid down' had . . . or has since acquired a meaning in electronic or semiconductor arts which necessarily connotes adherence." In ignoring the difference between the crucial phrases, the appeals court said, the interference board was "clearly in error." The board's opinion was reversed. The Borovoy ploy had worked. Once again, Robert Noyce was officially recognized as the inventor of the microchip.

Now it was Mosher's turn to appeal. Six months after the court of appeals' opinion, he filed a brief in the U.S. Supreme Court, asking the justices to review the opinion. Six months later, the Court issued a terse reply to Mosher's request: "Denied." Ten years and ten months after Jack Kilby had first applied for his patent, the case of *Kilby* v. *Noyce* had come to an end. Noyce had won. Now the stage was set for Fairchild to exploit its patent.

Actually, it was not set. During the decade that the lawyers had been waging their battle, the integrated circuit had emerged as the most important new product in the history of electronics. The market grew explosively. By the time the last court had issued the last ruling, production of semiconductor chips was a multibillion-dollar industry. As a result, the legal right to this invention had become too important to be left to lawyers.

And so, in the summer of 1966, before the first opinion was issued, executives from Texas Instruments, Fairchild, and about a dozen other electronics firms had held a summit meeting and cut a deal. TI and Fairchild each conceded that the other had some right to the historic invention. The two companies agreed to grant licenses to each other for integrated circuit production. Any other firm that wanted to enter the market then had to arrange separate licenses with both Texas Instruments and Fairchild. The two firms generally demanded a royalty fee ranging from 2 to 4 percent of the licensee's profit from chip production. This agreement provided the other firms a means to enter the integrated circuit business; it has provided TI, Fairchild, and Fairchild's successor companies with hundreds of millions of dollars in royalties over the years.

Meanwhile, there was yet another court that had to weigh in—the court of professional opinion. That ruling was arguably more important than anything the Supreme Court could say about patent rights to the microchip. Here was an invention, after all, of transcendent importance, and it could clearly be traced to one of two specific men. That was unusual. Most contemporary innovations seem to emerge from vast, faceless research labs or from giant corporate R&D operations. The ideas tend to have a shared parentage that extends over a broad range of individuals. It's hard to name, for example, the inventor of the cellular phone, or the Internet, or AZT, global positioning satellites, digital cameras, artificial knees, vitamin pills, Viagra. (Actually, Pfizer Pharmaceuticals did file patents for Viagra, naming five specific scientists as the inventors. But Pfizer now says those five were merely representative of the vast research team that developed the penile dysfunction pill.)

The microchip was different. Two young men had independently hit on the monolithic idea at about the same time. Both worked in corporate settings that made it possible to bring the idea to fruition quickly. Because both had been trained to record and date their ideas in lab notebooks, there was documentary proof of their inventions. So who would get the credit? Which one would go down in history as the Man Who Made the Microchip?

The question could fairly easily have become an ugly and extended point of contention. In fact, it never did. This salutary result is largely due to the nature of the two inventors. Both were decent, fair-minded people. Both were men who got more pleasure from the sheer joy of inventing than from public acclaim for their inventions. Both Jack Kilby and Bob Noyce were far more comfortable at the lab table, working out some technical problem, than they were at the head table of some gala banquet held in their honor. Accordingly, almost from the beginning, both engineers were generous about recognizing the work of the other guy. Taking a cue from the two inventors, the scientific and engineering communities agreed to agree that Kilby and Noyce deserved joint credit for the monolithic idea.

Both Kilby and Noyce were awarded the National Medal of Science for overcoming the tyranny of numbers, and both were inducted into the National Inventors Hall of Fame as inventors of the integrated circuit. In the engineering textbooks, Kilby gets credit for the idea of integrating components on a chip, and Noyce for working out a practical way to connect those components. Among their fellow engineers, Kilby and Noyce are referred to as co-inventors of the chip, a term that both men found satisfactory. After Noyce's death in 1990, Jack Kilby kept getting awards, including the Kyoto Prize, the Japanese version of the Nobel Prize (in 1993) and then the Nobel Prize in Physics (in 2000). On both occasions, Kilby pointed out that "Bob Noyce of Fairchild developed a similar idea, along with a practical means of manufacturing it."

So the praise as well as the profit for this groundbreaking idea was to be shared. On the day of Fairchild's great courtroom victory, consequently, hardly anybody paid any attention. After ten years, tens of thousands of pages, and well over a million dollars in legal fees, the legal labors had brought forth an utterly inconsequential mouse. "Patent Appeals Court Finds for Noyce on IC's" began the headline over a small story reporting the decision in the trade journal *Electronic News*. "IC Patent Reversal Won't Change Much."

THE REAL MIRACLE

The integrated circuit made its debut before electronic society at the New York Coliseum on March 24, 1959. The occasion was the industry's most important yearly get-together—the annual convention of the Institute of Radio Engineers. Texas Instruments had managed, in the nick of time, to turn out a few chips that had no flying wires, and there was a lavish display at the TI booth featuring the new "solid circuits." There was also a lavish prediction (which we know today to have been a massive understatement) from TI's president, who said that Jack Kilby's invention would prove to be the most important and must lucrative technological development since the silicon transistor. Nonetheless, the new circuit-on-a-chip received a frosty reception.

"It wasn't a sensation," Kilby recalls dryly. There were about 17,000 electronic products on display at the convention (the Coliseum used a million watts of power daily during the gathering), and large numbers of them attracted more attention than the integrated circuit. There were hundreds of reporters on hand, and virtually all of them managed to miss the biggest story of the week. In its special issue on the convention, *Electronics* magazine, which was supposed to recognize important new developments in the field, offered breathless reports on such innovations as a backward-wave oscillator and a gallium arsenide diode, but made no mention of the integrated circuit. In a wrap-up two weeks later, *Electronics* devoted a single paragraph to Texas Instruments' new "match-head size solid-state circuit."

"There was a lot of flak at first," Kilby recalls, and indeed, what little comment the new device received was largely critical. The critics identified three basic problems with the integrated circuit. In the first place, the idea of making resistors and capacitors out of silicon flew in the face of decades of research that had established conclusively that nichrome was the optimum material for making resistors, Mylar for capacitors. Monolithic circuits of silicon would be inherently inferior. In the second place, integrated circuits would be hard to make; one common line of analysis held that 90 percent of each production batch of chips would be faulty. In the third place, the whole concept posed a threat to an important segment of the engineering community. If component manufacturers like Texas Instruments started selling complete circuits to computer manufacturers, the circuit designers at computer firms would become redundant—and unemployed.

"These objections were difficult to overcome," Kilby wrote later, "because they were all true." As a result, the giants of the industry—Sylvania, Westinghouse, and their ilk—carefully kept themselves clear of the business for several years. This untimely burst of caution opened the way for upstarts like Texas Instruments, Fairchild, and a slew of new firms in Silicon Valley to work out the problems and cash in on the revolution. With intensive research, the hungry young companies learned how to design circuits on the chips that circumvented the shortcomings of silicon components; they found new production techniques that overcame the initial manufacturing difficulties. The result has been American industry's greatest postwar triumph. The integrated circuit, a child of Texas and California, has swept the world and spawned a furiously competitive global market. At the start of the twenty-first century, annual sales of integrated circuits were close to $200 billion; the market for digital devices dependent on the chip was well over $1 trillion per year. Just a few years after its unspectacular coming-out party, the integrated circuit caught the attention of the press and became known as the "miracle chip." Today, the miraculous has become normal, and the chip is ubiquitous. The average home in any developed country contains thousands of integrated circuits; the average garage has hundreds more. The 24,800 man-made objects presently floating in space are crammed with millions of integrated circuits and would not be up there if they weren't.

The integrated circuit was an enormous success because it solved an enormously important problem—the tyranny of numbers. But the success story was also a matter of timing. The chip was born just when the computer was starting to grow up. When Kilby and Noyce made their intellectual breakthrough, the computer was right on the verge of becoming an essential tool for agencies and companies with major number-crunching requirements—banks, insurance companies, the Social Security system, etc. But even in the 1950s, a few visionaries were talking about the concept of a "personal" computer—a computer in every home, or a computer on every wrist. Chips were perfect tools to implement the digital math and logic that computers use, and they were small enough to permit the computer to shrink without losing capacity. The chip and the computer went together like the horse and carriage—or, more aptly, like the oil industry and the auto industry that sprang up together at the start of the twentieth century. "The synergy between a new component and a new application generated an explosive growth for both," Bob Noyce wrote in a retrospective article two decades after the monolithic idea was born. "The computer was the ideal market . . . a much larger market than could have been provided by the traditional applications of electronics in communications."

In traditional circuitry, involving discrete components wired together, resistors and capacitors were cheap, but switching components such as vacuum tubes and transistors were relatively expensive. This situation was nicely suited to the manufacture of radios, television sets, and the like; an ordinary table radio of the 1950s used two or three dozen capacitors and resistors but only two or three transistors. With integrated circuits, the traditional economies were reversed. Resistors and capacitors, which use up power and take up a lot of room on a chip, became expensive; transistors were compact, simple to make, and cheap. That situation is precisely suited for computers and other digital devices, which need large numbers of switches—transistors—and small quantities of other components.

The innards of computers, calculators, cell phones, digital cameras, etc., consist of chips containing long chains of transistors that switch back and forth to manipulate information. Like a light switch on the wall, these transistors can be either on or off; there's nothing in between. Since there are only two possible conditions, a computer has to reduce every job, every decision, every computation to the simplest possible terms: on or off, yes or no, stop or go, one or zero. Humans can do the same thing, of course. We do it on Easter morning when the kids look for hidden eggs and their parents provide only two clues: "You're hot" or "You're cold." Eventually, most of the eggs are found, but the process is so tedious that even the kids get fed up with it pretty quickly. Computers, in contrast, use this tedious system all day, every day. They have to. They can't handle anything else.

For all the mystique of "electronic brains" and "artificial intelligence," digital devices are actually mindless dullards that rely on computational techniques mankind abandoned in Neanderthal days. Digital problem solving involves simple math—far simpler than the stuff humans learn in grade school. A computer approaches every problem like a child counting on his fingers, but the computer counts as if it had only one finger. (The word "digital" comes from the Latin *digitus*, meaning "a finger.") The real miracle of the "miracle chip" is that people have devised ways to manipulate this one minimal skill so that machines can carry out complex operations.

Although digital devices can recognize only two numbers—1 and 0—they can arrange those two to represent any number in the universe. Large computers have, for example, computed the value of pi to 206 trillion decimal places—a number so long it would take a 4-million-page book just to print it. But the only numbers the computers had to work with were 0 and 1.

To understand how this works, forget numbers for a moment and think about words—specifically, the word "cat." Anyone familiar with languages knows that there is nothing inherent in the purring, four-footed feline species that requires it to be represented by the three Roman letters "C," "A," and "T." You can spell the thing *gato* or *chatte* and it is still the same animal. For that matter, there's no reason why a cat has to be represented by letters chosen from our 26-letter Roman alphabet. In the story *On Beyond Zebra*, Dr. Seuss tells of a boy who decided that the Roman alphabet was incomplete, so he went on past Z and invented 19 new letters, including one, "thnad," that is used to spell the name of a typically Seussian breed of cat. An alphabet can have as many letters in as many different shapes as its users find convenient. When Samuel F.B. Morse perfected his telegraph, he found that only two letters—dot and dash—could be conveniently sent through the wires. With those two, he invented his own alphabet; telegraphers used to practice their skill by tapping out the Morse Code version of "cat": ·—· ·——. The Japanese kana alphabet has 47 different letters; in that system "cat" is represented by the symbols ね こ, but it means the same thing as our word "cat." The Chinese alphabet includes about 25,000 different characters, including one that stands for "cat": 猫. The 26-letter alphabet we use is just a selection of symbols that Western culture has grown used to. The word "cat" is just a convenient combination of those symbols that English speakers have settled on to represent a cat.

The same principles apply to numbers. There's nothing inherent in the number 206, just to choose a number at random, that requires it to be represented by the symbols 2, 0, and 6. We just happen to use that representation because of the way our number system works. Our system uses ten different symbols, or digits, for numbers:

$$0\ 1\ 2\ 3\ 4\ 5\ 6\ 7\ 8\ 9$$

To represent numbers larger than 9, we add a second column and run through the list again:

$$10\ 11\ 12$$

Since the turning point—the new column—comes at the number 10, our system is called the base-10, or decimal, number system. The decimal system is the most familiar counting system in the world today, but it is by no means the only one available. It would be quite simple to go On Beyond Nine and invent new symbols; a duodecimal, or base-12, number system, could look like this:

$$0\ 1\ 2\ 3\ 4\ 5\ 6\ 7\ 8\ 9\ \&\ \#\ 10$$

With twelve different symbols, the turning point—the new column—comes at the number 12.

The great breakthrough that permitted man to count far beyond 10 with just ten different symbols was the invention of this turning point—a concept that mathematicians call positional notation. Positional notation means that each digit in a number has a particular value based on its position. In a decimal number, the first (farthest right) digit represents 1's, the next digit 10's, the next 100's, and so on. The number 206 stands for six 1's, no 10's, and two 100's:

$$100 \quad 10 \quad 1$$
$$\times 2 \quad \times 0 \quad \times 6$$

Add it all up:

$$200 \; + \; 0 \; + \; 6$$

and you get 206. This number, incidentally, demonstrates why mathematicians consider the invention of a symbol that represents nothing (i.e., the number 0) to have been a revolutionary event in man's intellectual history. Without zero, there would be no positional notation, because there would be no difference between 26 and 206 and 2,000,006. The Romans, for all their other achievements, never hit on the idea of zero and thus were stuck with a cumbersome system of M's, C's, X's, and I's which made higher math just about impossible.

With positional notation, we can use any number of different symbols to count with. We could devise a numerical alphabet with 26 different digits, or 206 different digits, or 2,006 different digits. The base-10 system we use is just a convenient method that people have settled on to represent all numbers.

If you want to know why modern man has settled on a base-10 number system, just spread your hands and count the digits. All creatures develop a number system based on their basic counting equipment; for us, that means our ten fingers. The Mayans, who went around barefoot, used a base-20 (vigesimal) number system; their calendars employ twenty different digits. The ancient Babylonians, who counted on their two arms as well as their ten fingers, devised a base-12 number system that still lives today in the methods we use to tell time and buy eggs. Someday a diligent grad student doing interdisciplinary work in mathematics and the history of film may produce a dissertation demonstrating that the residents of E.T.'s planet use an octal number system; the movie shows plainly that E.T. has eight fingers. For earthbound humans, however, the handy counting system is base-10.

A computer's basic counting equipment is simpler. It is an electronic switch—a transistor—that can be either on or off. Each of these conditions represents one digit; on represents 1 and off represents 0. This two-digit number system is called the base-2, or binary, system. Just as people can count to any number, no matter how high, with just ten digits, a computer can count to any number with just two. Like people, computers do this through positional notation. Counting in binary starts out just like decimal:

$$0 \quad 1$$

But in a binary number, the turning point—the new column—comes at the number 2. In binary the two-digit number 10 stands for one 2 and no 1's—that is, the quantity 2. Binary 11 means one 2 plus one 1—that is, 3. Another column must be added to write 4, another for 8:

$$0 \quad 1 \quad 10 \quad 11 \quad 100 \quad 101 \quad 110 \quad 111 \quad 1000$$

Things go on this way until we get to the number 1111. Reading from the right (generally the easiest way to read a binary number), 1111 stands for one 1 plus one 2 plus one 4 plus one 8, or 15. Having come to 1111, the system is out of digits again, so another column is added: the number 10000 is the binary version of 16.

In a binary number, in other words, the first (far right) column represents 1's, the second column 2's, the third 4's, the fourth 8's, the fifth 16's, and so on, as long as necessary. The binary number 11001110, just to choose a number at random, represents (from the right), no 1's, one 2, one 4, one 8, no 16's, no 32's, one 64, and one 128:

128	64	32	16	8	4	2	1
×1	×1	×0	×0	×1	×1	×1	×0

Add it all up

$$128 + 64 + 0 + 0 + 8 + 4 + 2 + 0 = 206$$

and it turns out that 11001110 is precisely the same number that is represented by 206 in the decimal system. The quantity hasn't changed; the only thing different is the alphabet, or number system, used to represent it.

Except for someone who has lost nine fingers in an accident, a number system based solely on 1 and 0 is not particularly useful for humans. But the binary system is perfect for digital machines, not only because all numbers can be represented by chains of on and off electronic switches, but also because binary arithmetic is the simplest possible mathematical system. Humans learning basic arithmetic in the elementary grades find that our decimal system requires them to memorize hundreds of "math facts" — facts like $2 + 2 = 4$, $17 - 9 = 8$, $8 \times 7 = 56$. If they were learning the binary system, things would be infinitely simpler. In binary arithmetic, there are only three math facts: $0 + 0 = 0$, $1 + 0 = 1$, and $1 + 1 = 10$ (10, of course, is the binary version of decimal 2).

This extreme simplicity is a boon to computer designers. It is possible, but cumbersome, to construct extensive mazes of transistors inside a machine to perform binary subtraction, multiplication, and division. Thanks to some convenient mathematical gimmickry, however, digital machines based on the binary system can carry out any numerical operation using only addition, which involves fairly simple circuitry. With an ancient trick called ones-complement subtraction, computers can solve subtraction problems by adding. Multiplication is performed the way humans did it eons ago, before they developed the multiplication table—by repeated addition. If you ask your calculator to multiply 4×1000, the machine steers the numbers through a series of logic gates that essentially put a binary 4 in a storage place, and then 999 more 4's, one at a time, to get the answer. Division, similarly, becomes a series of ones-complement subtractions.

The ingenious, indeed breathtaking, insight that binary mathematics was perfectly suited to electronic computers occurred more or less simultaneously on both sides of the Atlantic to a pair of ingenious, indeed breathtaking, visionaries who had scoped out, by the late 1940s, remarkably accurate forecasts of the development of digital computers over the ensuing half century. These two cybernetic pioneers were John von Neumann and Alan M. Turing.

Von Neumann was born in Budapest, the son of a wealthy banker, in 1903. He was recognized almost immediately as a prodigious mathematical talent, and spent his youth shuttling from one great university to another: Berlin, Zurich, Budapest, Göttingen, Hamburg. He published his first scholarly monograph at the age of eighteen and thereafter turned out key papers in a wide variety of fields. In 1930 he sailed west with a tide of refugee European scholars to Princeton, where he held a chair at the university but also became one of the first fellows—along with a fellow named Albert Einstein—of the Institute for Advanced Study. He made important contributions in pure mathematics, but also wrote major works on applications, ranging from chemical engineering and quantum physics to economics and the Theory of Games, a mathematical construct of his own for winning complex games.

During World War II, von Neumann was involved with the development of atomic bombs, an engineering task of overwhelming scope that required, among much else, huge numbers of separate mathematical computations. On a train platform in Aberdeen, Maryland, one day in 1944, von Neumann was pondering how best to scale this mountain of mathematics when, by sheer luck, he ran into a younger mathematician, Herman Goldstine. Chatting on the train, Goldstine told von Neumann about the new ENIAC computer under way in Philadelphia—a machine that could zip through repetitive computations at unprecedented speed. This chance conversation pulled von Neumann into the new world of computers, where he immediately began making major contributions.

A lot of people who visited ENIAC in the postwar years saw an amazing leviathan of hot, blinking tubes that could, indeed, handle huge computational problems but had no other practical use. When the farsighted von Neumann looked on ENIAC for the first time, though, he saw a future in which computing machines were universally used, with universal applications. The transistor, with its promise of fast, low-power switching, spurred him to even more ambitious theories of what computers might become. More and more, toward the end of his life, he began to see parallels between the evolution of computing machines and the evolution of the human mind. His last book, published posthumously in 1958, was titled *The Computer and the Brain*.

Alan Turing, born in London in 1912, was considered a poor student with little academic promise through most of his school career. After twice failing the scholarship exam for Trinity College, Cambridge, he matriculated at King's, another Cambridge college, and took his Ph.D. there in 1935. He became intrigued by the Entscheidungsproblem, a deep mathematical quandary posed by the German scholar David Hilbert. One difficulty in solving this problem was that a solution would take not only ingenuity but also drudgery, because it required endless repetitive calculations. But endless repetitive chores were a waste of human time and energy, Turing felt; the drudgery should be left to machines. While pondering Hilbert's problem, Turing hit upon an extraordinary new idea: that a machine could be designed, or programmed, to perform any mathematical computation a human could carry out as long as there was a clear set of instructions for this machine to follow. This ideal computer, with universal application as long as it was programmed correctly, came to be called the Turing Machine, and the concept served as a key inspiration for computer pioneers in Europe and the United States—among them von Neumann, whom Turing met during a stay at Princeton in the mid-thirties.

During the war, Turing joined the team of mathematicians who gave the Allies an invaluable step up by cracking the Germans' Enigma military code. The work involved reading pages and pages of sheer gobbledygook, looking for repetitive patterns of letters that would reveal, under ingenious mathematical manipulation, the inner workings of the German cipher machines. To carry out the calculations, the codebreakers developed simple mathematical machines of their own—real-life variations on the abstract Turing Machine. After the war, Turing worked on the first generation of British computers. At the age of forty-one, shortly after being tried and convicted for homosexual conduct ("Accused Had Powerful Brain," a London tabloid reported), he died from eating an apple tainted with cyanide he was using for an experiment. People are still debating today whether that was an accident.

Turing had gone even further than von Neumann in suggesting that electronic "brains" could eventually match those of their human builders. "One day ladies will take their computers for walks in the park," he predicted, "and tell each other, 'My little computer said such a funny thing this morning.'" In his most famous paper, "Can a Machine Think?" published in 1951, he predicted that computers would be carrying on "human" conversations with men and with other machines "in about fifty years' time." The filmmaker Stanley Kubrick read that monograph, did the addition, and went to work on a movie about the year 2001.

Von Neumann, in a report to the U.S. Army Ordnance Department, and Turing, in a report for England's National Physical Laboratory, set forth their notions of the general architecture of an electronic computer. Both agreed that the device would have to carry out four basic functions: "input," to take in data and instructions; "memory" (in England, "store"), to keep track of the data; "processing," to do the actual computing; and "output," to report the answer back to the human user. And both concluded that the logical way to handle data was in the form of binary numbers.

Since computer calculations would be performed by switches flipping from on to off and back again, Turing wrote, it was natural enough to assign the value 1 to on and 0 to off, and handle all mathematics with only those two digits. "We feel strongly in favor of the binary system for our device," the von Neumann report agreed. The computer "is naturally adapted to the binary system since we . . . are content to distinguish [just] two states," he wrote. "The main virtue of the binary system . . . is, however, the greater simplicity and speed with which the elementary operations can be performed."

For all the virtues of binary, though, there was a problem— something the mathematicians called the "conversion problem." This was a euphemism for the fact that few humans understand the binary system and thus would find it difficult to convert a computer's answers into a form intelligible to people.

Two solutions were proposed. One was that the human race should drop its decimal system and learn something closer to binary. Under this arrangement, preschoolers watching *Sesame Street* would be indoctrinated with rhymes like this:

1   10
Buckle my shoe

11   100
Shut the door,

101   110
Pick up sticks

111   1000
Lay them straight

1001   1010
The big fat hen.

Such a sharp change in human habits was obviated by a more practicable idea set forth by both Turing and von Neumann. "The one disadvantage of the binary system from the human point of view," von Neumann's report noted, "is the conversion problem. Since, however, it is completely known how to convert numbers from one base to another . . . there is no reason why the computer itself cannot carry out this conversion."

This suggestion was quickly adopted, and ever since all digital devices have included a piece of circuitry called a decoder, which translates decimal numbers into their binary equivalents. When you punch the keys to put the number 206 into your calculator, the decoder sends out electronic pulses to a chain of eight transistors so that the transistors line up this way:

ON	ON	OFF	OFF	ON	ON	ON	OFF
1	1	0	0	1	1	1	0

Thus transformed to binary format—11001110—the number 206 becomes comprehensible to a digital machine.

It was human genius on the part of von Neumann, Turing, and others like them that figured out how to use binary numbers and binary math to turn an inert chain of electronic switches into a powerful computational tool. But the computer pioneers did not stop there. They also designed a complete system of logic that permits machines to make decisions and comparisons and thus work through complex patterns, or "programs," for manipulating words and numbers. The beauty of this logic system is that it, too, is binary; it can be implemented by integrated circuits full of transistors that do nothing but switch on and off.

Which modern high-tech genius developed this binary logic? None. The logical methods that all digital devices use today were worked out about 100 years B.C. (before computers) by a British mathematician named George Boole.

Boole was born in Lincolnshire in 1815, the son of a cobbler who was always pressed for money. The family's lowly status dictated that the boy would enter some manual trade; he was sent to a vocational school that did not even attempt to teach Latin, the sine qua non of a professional future for any English lad of that day. Undaunted, George taught himself Latin and Greek after school. This came in handy in 1831, when the sixteen-year-old boy was forced to leave school and help support the family. He took a job as an assistant teacher, but continued to educate himself. In his seventeenth year Boole had two experiences that changed his life. He read Newton's *Principia* and transferred his attention from classical languages to math. Shortly afterward, while walking alone through an open field, Boole was suddenly struck with a "flash of psychological insight" that convinced him that all human mental processes could be formulated in straightforward mathematical terms.

It would be pleasant to report that Boole then and there dedicated his life to the explication of this great concept. Unfortunately, things were not that easy. The family was now dependent on George for support, and his job left insufficient time for complex mathematical work. His knowledge of Latin and Greek qualified him for the clergy, and he decided to train for ordination. Gradually, though, it became clear that Boole was too much a freethinker for such a career; unlike the Anglican Church, he doubted the literal truth of the Bible and believed in religious tolerance. All his life, in fact, he was suspicious of clergymen and their efforts at indoctrination. On his deathbed, according to a biographer, William Kneale, Boole requested "that his children not be allowed to fall into the hands of those who were commonly thought religious."

Committed to teaching, Boole opened his own school in Lincolnshire and now found some time for mathematical work. The editor of a new journal was willing to publish Boole's papers, despite the author's lack of formal training. One of them caught the eye of the mathematician Augustus De Morgan, who helped Boole obtain a chair at Queen's College in Ireland. At last Boole had a secure income and the time to work out his grand mathematical synthesis of human thought. In 1854, after a decade of intense work, he published his masterwork, *The Laws of Thought, on Which Are Founded the Mathematical Theories of Logic and Probabilities.* "The design of the following treatise," the book begins, "is to investigate the fundamental laws of those operations of the mind by which reasoning is performed; to give expression to them in the symbolic language of a Calculus; and upon this foundation to establish the science of Logic . . ." Completely new, and somewhat obscure even to the expert, it had little initial impact. Today the book is recognized as a milestone that did indeed establish the new science of symbolic logic.

In Ireland, Boole married Mary Everest, niece of Sir George Everest, the geographer who surveyed the high mountains of Nepal and left his name to the highest of all. Throughout his life, he demonstrated the phenomenal energy that characterized the Victorian age. In addition to his work on logic, Boole published two widely used textbooks and countless monographs. He found time for poetry, including a difficult lyric entitled "Sonnet to the Number Three." He was a trustee of the Female Penitents' Home and an officer of the Early Closing Association, which strove to reduce the workday to ten hours. A photograph shows him to be an intense, thoughtful professor with a square face, dark hair, and penetrating eyes. For all his achievements in higher math, he never shirked his duties as a teacher. In November 1864 he walked two miles through a cold rain to meet a class and proceeded with the lecture in his sodden clothes. From this he contracted pneumonia and died. He left behind one last manuscript, so singular and so arcane that the experts at the Royal Society could not decipher it. "No mere mathematician can understand it," his widow observed, "and no theologian cares to try."

Since Boolean logic—also known as Boolean algebra, because Boole expressed logical concepts in algebraic terms—is now recognized as something important, the academicians have draped it in a formidable veil of complicated jargon, symbols, and formulas. At the core, though, the Laws of Thought that Boole described in mathematical terms are the stuff of everyday life. Boole examined everyday mental processes in terms of the simple connective tissue of language: and, or, not.

You wake up from a sound sleep. Can you roll over and sleep some more, or do you have to get up and go to work? To decide, you carry out a fundamental Boolean operation. If your clock says yes, it's after 8:00, and your calendar says yes, it's a weekday, then yes, you get up for work. If either of these conditions is a no, however, you can stay in bed. This decision is known today as a

Boolean AND operation. The result is yes only if condition 1 AND condition 2 are both yes.

The kitchen sink represents another basic Boolean pattern. If no faucet is on, no water comes out of the spigot. But if either the hot faucet OR the cold faucet is on, OR if both are on, water will flow. This decision, in which the result is yes if condition 1 OR condition 2 OR both are yes, is known as a Boolean OR operation.

In essence, Boole demonstrated that all human reasoning could be reduced to a series of yes-or-no decisions. Each decision could therefore be represented in algebraic terms. Sometimes the formulas were as simple as $x + y - z$, and sometimes they were more complex; in The Laws of Thought, Boole formulates an argument that God exists, as follows: $x(1 - y)(1 - z) + y(1 - x)(1 - z) + z(1 - x)(1 - y) = 1$. The most important of Boole's algebraic formulas—the one he describes as the central pillar of his entire yes-or-no structure—is this:

$$x = x^2$$

Anyone young enough to remember high school algebra will see that this equation holds true for two, and only two, numbers: 0 and 1. In other words, Boole's organization of all human decisions into yes-or-no terms turned out to be a binary system. A century ahead of time, the self-taught Victorian scholar had developed a decision-making methodology that would prove just right for digital machines.

Until digital machines came along, however, Boole's algebra was largely ignored, except by a few of his fellow logicians. One of Boole's most avid followers was the Oxford don in mathematics Charles Lutwidge Dodgson, who wrote a series of academic works on symbolic logic and who, under his pen name, Lewis Carroll, sprinkled his "Alice" books with allusions to Boole's ideas. Many of the people Alice meets beyond the looking glass see their world in basic Boolean terms—yes-or-no, true-or-false, does-or-doesn't:

> "You are sad," the Knight said in an anxious tone. "Let me sing you a song to comfort you."
>
> "Is it very long?" Alice asked, for she had heard a good deal of poetry that day.
>
> "It's long," said the Knight, "but it's very, *very* beautiful. Everybody that hears me sing it—either it brings the *tears* into their eyes, or else—"
>
> "Or else what?" . . .
>
> "Or else it doesn't, you know."

Bertrand Russell was another admirer. In *Principia Mathematica,* the Promethean effort to set down once and for all the fundamental logical basis of all mathematics, Russell and Alfred North Whitehead carried Boole's original concept to a climactic conclusion. Unstintingly meticulous (it takes the authors one and a half volumes to arrive at their proof that $1 + 1 = 2$) and inaccessible to all but a small coterie of experts, the Russell-Whitehead treatise seemed to offer further proof, if any were needed, that Boole's curious combination of logic and algebra was an intellectual abstraction devoid of practical use. This was hardly what Boole had intended: "The abstract doctrines of science," he writes in *The Laws of Thought,* "should minister to more than intellectual gratification." Fifty years after it appeared, though, Boole's great work was considered strictly an academic exercise.

The narrative now shifts ahead to 1937 and across the Atlantic to Cambridge, Massachusetts, where groups of engineers and mathematicians were struggling to design the first primitive versions of a digital computer. An MIT engineer, Vannevar Bush, had designed an electrical calculating machine that used decimal numbers; it was built of rods, shafts, and gears arranged so that a gear would turn one tenth of a full rotation (36 degrees) to represent the number 1, two tenths (72 degrees) for 2, and so on. This device, although revolutionary for its day, tended to be imprecise; if the gear happened to turn 48 degrees, or 55 degrees, what number was represented? And so attention shifted down the street to Harvard, where another engineer, Howard Aiken, was thinking—as Von Neumann and Turing had been—about a binary machine that would use simple electrical switches. On the binary computer, precision was not a problem—the switches were either on or off, nothing else—but it was a forbiddingly complicated task to design the proper combinations of switches to carry out binary arithmetic.

An MIT graduate student, Claude E. Shannon, who had been working with Bush, was looking for a thesis topic and decided to take on the important but formidable problem of designing digital switching circuits. In the course of his work, Shannon hit upon a crucial idea.

If society allocated fame and fortune on the basis of intellectual merit, Claude Shannon would have been as rich and as famous as any rock idol or football star. Born in the farm community of Gaylord, Michigan, in 1916, he graduated from the University of Michigan in 1936 and went on to take a Ph.D. in electrical engineering at MIT. His master's thesis, in 1937, demonstrated how computerized mathematical circuits should be designed; this youthful piece of work not only served as the cornerstone of computer architecture from then on, but also launched a new academic discipline known as switching theory. Ten years later, as a researcher at Bell Labs, Shannon got to thinking about efficient means of electronic communications (for example, how to send the largest number of telephone conversations through a single wire). He published another seminal paper, "A Mathematical Theory of Communication," that launched an even more important new academic discipline known as information theory; today information theory is fundamental not only in electronics and computer science but also in linguistics, sociology, and numerous other fields. You could argue that Claude Shannon was the Alexander Graham Bell of the cellular phone, because mobile communications would be impossible without the basic formulas of information theory that Shannon devised.

In 1949, Shannon published a monograph—once again, the first one ever written on the topic—called "Programming a Computer for Playing Chess." The ideas set forth there are still central to the design of all computer games, including Deep Blue, the program that defeated world chess champion Garry Kasparov. Like many mathematicians, Shannon was an avid fan of games and puzzles; among other things, he liked to work out "pangrams"—sentences that contain every letter of the alphabet. His *pièce de résistance* in this field is a sentence that uses each letter only once: "Squdgy fez, blank jimp crwth vox!"

That fascination with language was useful for a man working at the very front edge of technology. New concepts required new terms, and Shannon made several contributions to the language of high-tech. Looking for a word to mean a single unit of information—in digital terms, a 1 or a 0—he first coined the phrase "binary digit" but quickly shortened that to "bit." To this day, the capacity of computers and other digital devices is still measured in bits; if a personal computer is rated at 64 megabits, that means it comes with enough random-access memory to store 64 million bits, or distinct pieces of information. The term is now used by digital designers everywhere, many of whom have probably never heard of Claude Shannon. Shannon wouldn't mind that, though. He was not one to blow his own horn. During the years he taught information theory at MIT, he never mentioned that he was the creator of the academic discipline his students were studying, and seemed somewhat embarrassed when diligent students figured out that their prof was the progenitor. Early in 2001, Bell Labs set up an exhibit in Shannon's honor, noting how many of his twentieth-century ideas have become part and parcel of daily life in the new century. Shannon stayed away from the opening ceremony. A few weeks later he died, receiving brief obituaries in a few papers. Hardly anybody seemed to remember how influential he had been in shaping the modern world.

In 1937, when Shannon tackled the problem of binary circuit design, digital computers used magnetic switches called relays. A relay looks like a mousetrap with an electromagnet on one end. When electricity flows to the magnet, it attracts the metal bar of the mousetrap, which flips over and thus turns the switch on. As soon as the current is cut off, the magnetic attraction stops and the metal bar springs back, turning the relay off. The problem was how to design arrays of these relays so that they would switch on and off in the proper order to add binary numbers. This was seen as an inordinately difficult task, requiring the designer to contemplate so many different levels of possible variations that it would tend to drive anybody crazy. Today, the tedious, repetitive work of designing computer architecture is a chore left to computers. But in 1937 there weren't any. It would be up to people to figure out the design of binary logic.

While stewing over this question, Shannon happened upon a text on Boolean logic and something clicked. Boole's equations for AND operations, OR operations, and other logical functions reduced decision making to a set of dualities—yes or no, 0 or 1, true or false. Shannon recognized that these pairs could be represented just as well by the switching duality: on or off. In short, the dreadfully formidable task of designing binary logic circuits had already been done—by George Boole. Boole's carefully worked-out equations could serve as road maps for wiring together electric switches to carry out logical operations. Accordingly, Shannon wrote, "It is possible to perform complex mathematical operations by means of relay circuits. Numbers may be represented by the positions of relays and stepping switches, and interconnections between sets of relays can be made to represent various mathematical operations." At the end of his paper, Shannon showed how a series of relays, arranged to carry out Boolean AND and OR operations, could be wired to add two binary numbers.

In addition to mathematical operations, Shannon demonstrated, Boolean circuits could be wired to make comparisons—is number x equal to number y?—and to follow simple directions of the "If A, then B" category. "In fact, any operation that can be completely described in a finite number of steps using the words 'if,' 'or,' 'and,' etc.," Shannon wrote, "can be done automatically with relays." With this ability to make decisions—to proceed in different ways depending on the results of its calculations—the machine could be programmed to carry out complicated computations without constant direction from the human operator.

The techniques set forth in Shannon's thesis have been universally adopted for digital machines. In modern computers transistors embedded in integrated circuits have replaced magnetic relays, but the principles of binary switching remain the same. A transistor built into integrated circuits is essentially the same silicon sandwich developed by Shockley's team: it consists of a thin layer of P-type silicon sandwiched between two slightly thicker layers of N-type silicon. The device is hooked up so that current— a surge of electrons—will run from one N-type layer, through the middle, and out the other N-type layer. This current flow is switched on and off by signal pulses flowing to the middle layer. If a pulse is sent to the center layer, current will flow from end to end; the transistor is on. But if the center receives no pulse, it blocks current flow from N to N; then the switch is off.

A computer's circuitry is a chain of transistors, one after another. The circuit is analogous to a long irrigation pipe with a series of faucets built into it to control the flow of water. If faucet 1 is open, water can flow along to faucet 2; if that one is open, water can flow on to faucet 3. By opening the right combination of

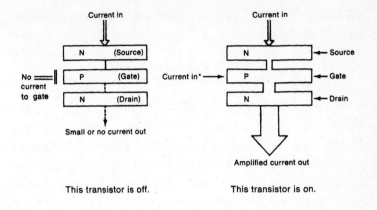

This transistor is off.　　　　This transistor is on.

* The greater the current flowing into the gate, the greater the flow from source to drain.

faucets at the right time, a farmer can direct water to any point in his field. In a computer's electronic pipeline, each transistor acts as a faucet; if transistor 1 is on, current can flow on through to transistor 2, and so on. By turning on the right combination of transistors at the right time, computer designers can direct the flow of current to any point in the circuit.

To set up the right combinations, computer builders rely mainly on three basic Boolean circuits. The simplest AND circuit, in accordance with Boole's AND operation, consists of three transistors lined up so that switch 3 is on only if both switch 1 AND switch 2 are on. The arrangement looks like this:

If either switch 1 or switch 2 is off, the flow of current will be blocked and switch 3 must be off. Only if both 1 AND 2 are on will current flow through to turn on 3.

A simple OR circuit can be implemented with three switches arranged this way:

If both switch 1 and switch 2 are off, current flow will be blocked and switch 3 must be off. But in this circuit, current can flow through either switch 1 or switch 2 to get to switch 3. Thus if either 1 OR 2 is on, OR if both are on, current will flow through the circuit to turn on 3.

The third basic circuit, called a NOT circuit, can be wired from two switches arranged so that the second is NOT in the same state as the first. If switch 1 is on, 2 is off; if 1 turns off, 2 switches on.

Because these switch arrangements serve either to block current or to let it pass, they are commonly called gates. Logical and mathematical operations are carried out by sending current through a maze of different gates. A basic addition circuit, in its simplest form, can be implemented with a dozen AND gates, a half dozen OR gates, and three NOT gates. Pulses representing the binary numbers to be added are sent into the circuit. Each of these pulses turns selected transistors on or off in just the right combination so that the pulses coming out at the end of the circuit will represent, in binary, the sum of the two numbers that went in.

Mathematicians like to use the term "elegant" to describe a simple solution to a complex problem. The Boolean logic that adds two binary digits is the height of elegance. In the early computers, though, the electronics of this elegant arrangement were extremely cumbersome. The simple addition circuit just described, with its twenty-one separate gates, requires about fifty transistors, a dozen or so other components, and a labyrinthine spaghetti of connecting wires. And this circuit can add only two binary digits. If the numbers being added require, say, eight binary digits each—like the decimal number 206, which is 11001110 in binary—the problem would require eight separate passes through the addition circuit, plus a few dozen more transistors to store the result of each pass.

This is why computers were so vulnerable to the tyranny of numbers. The use of binary numbers and binary logic provided a precise computational system that was perfectly suited to electronic devices. But this perfect fit came at a price. The price was complexity. Digital devices are nothing more than switches turning on and turning off, but many, many switches must turn on and off many, many times to perform even simple operations. When electronic circuits had to be hand-wired together from individual components, these large numbers took an enormous toll in size, in speed, in cost, in power consumption, in difficulty of design.

Despite its inauspicious debut at the electronics convention, accordingly, the monolithic idea, conceived just as the digital computer was growing up, was destined to be a spectacular success. With integrated circuitry, the neat patterns of Boolean logic could be mapped directly onto the surface of a silicon chip; an entire addition circuit would now take up less space and consume less power than a single transistor did in the days of discrete components. With the advent of the chip, the digital computer had finally become as elegant in practice as it was on paper.

BLASTING OFF

The first integrated circuits proved so hard to produce that nearly two years passed after the chip's public debut before the new device was available for sale. Fairchild was first off the block; its catalogue for spring 1961 trumpeted a new line of six different monolithic circuits which it called "Micrologic elements." A few weeks later Texas Instruments entered the fray with a similar series of "solid circuits." As the two companies rather stridently pointed out, the new circuit-on-a-chip was smaller, lighter, faster, more power efficient, and more reliable than any conventional circuit wired together from discrete parts.

It was also more expensive. A "Micrologic" logic gate circuit, containing three or four transistors and another half dozen diodes and resistors, was initially priced at $120. An equipment manufacturer could wire together a circuit using top-of-the-line transistors for less than that, even after labor costs were figured in. It was as if an automobile company had designed a family station wagon that could go 500 miles per hour—and cost $150,000. Who needed it? "There was the natural reluctance to commit to something new," Bob Noyce recalled later. "And added to that you had a price that was basically uneconomical. So at first the traditional electronics customers just weren't buying."

This posed a fairly serious problem. Even more than most industries, electronics firms rely heavily on an economic phenomenon known as the learning curve. In the early life of a new product—when manufacturers are still learning how to design and produce the device at a reasonable cost—prices are necessarily high. As sales increase, better production techniques are developed, and prices curve sharply downward. The integrated circuit in early 1961 was stalled at the high end of the curve; there was no commercial market to push it down. As Noyce recalled, the chip seemed to be caught in a classic commercial Catch-22. Until the market picked up, the price would remain high; but as long as prices stayed high, the traditional electronics markets weren't interested.

And then, virtually overnight, the President of the United States created a new market.

In May 1961—a time when, as *The New York Times* noted, "there was a strong catch-the-Russians mood in Washington"— John F. Kennedy went before a joint session of Congress to propose "an extraordinary challenge." "I believe we should go to the moon," the president said. ". . . I believe that this nation should commit itself to achieving the goal, before this decade is out, of landing a man on the moon and returning him safely to earth. No single space project in this period will be more impressive to mankind. . . . And none will be so difficult or expensive to accomplish."

A 480,000-mile round trip to the moon was indeed a challenge of extraordinary dimensions for a nation whose greatest space achievement so far had been Alan Shepard's 15-minute, 302-mile suborbital flight in the spacecraft *Freedom* 7. (John Glenn, the third American in space, did not make his three-orbit trip until nine months after Kennedy's speech.) A successful lunar voyage would require major advances in rocketry, metallurgy, communications, and other fields. Among the most difficult problems were those the space experts called G & N—that is, guidance and navigation.

The trick of steering a fast-moving spaceship from a fast-moving planet through two different atmospheres and two different gravitational fields to a precise landing on a fast-moving satellite would require an endless series of instantly updated calculations—the kind of work only a computer could do. But a computer on a spacecraft would have to be smaller, lighter, faster, more power-efficient, and more reliable than any computer in existence. In short, somebody needed that 500-mph station wagon. And with the prestige of the nation at stake, high prices were no problem. "The space program badly needed the things that an integrated circuit could provide," Kilby said later. "They needed it so badly they were willing to pay two times or three times the price of a standard circuit to get it."

The G & N system was so crucial to the moon shot that the assignment for its development was the first major prime contract awarded after Kennedy's speech. There was no question that the computer would have to be built from integrated circuits, and Fairchild quickly started receiving large orders for its Micrologic chips. By the time the *Eagle* had landed at Tranquillity Base on July 20, 1969—meeting the late president's challenge with five months to spare—the Apollo program had purchased more than a million integrated circuits.

Two decades later, when the American semiconductor industry was facing an all-out battle with Japanese competitors, U.S. electronics companies complained loudly that Japanese firms had an unfair advantage because much of their development funds were provided by the government in Tokyo. On this point, the American manufacturers lived in glass houses. The government in Washington—specifically, the National Aeronautics and Space Administration and the Defense Department—played a crucial role in the development of the American semiconductor industry.

The Apollo project was the most glamorous early application of the chip, but there were numerous other rocket and weapons programs that provided research funds and, more important, large markets when the chip was still too expensive to compete against traditional circuits in civilian applications. A study published in 1977 reported that the government provided just under half of all the research and development money spent by the U.S. electronics industry in the first sixteen years of the chip's existence. Government sales constituted 100 percent of the market for integrated circuits until 1964, and the federal government remained the largest buyer of chips for several years after that.

The military had started funding research on new types of electric circuits in the early 1950s, when the tyranny of numbers first emerged. The problems inherent in complex circuits containing large numbers of individual components were particularly severe in defense applications. Such circuits tended to be big and heavy, but the services needed equipment that was light and portable. "The general rule of thumb in a missile was that one extra pound of payload cost $100,000 worth of extra fuel," Noyce recalled. "The shipping cost of sending up a 50-pound computer was too high even for the Pentagon." Further, space-age weapons had to be absolutely reliable—a goal that was inordinately difficult to achieve in a circuit with several thousand components and several thousand hand-soldered connections. When the Air Force ordered electronic equipment for the Minuteman I, the first modern intercontinental ballistic missile, specifications called for every single component—not just every radio but every transistor and every resistor in every radio—to have its own individual progress chart on which production, installation, checking, and rechecking could be recorded. Testing, retesting, and re-retesting more than doubled the cost of each electronic part.

In classic fashion, the three military services went off in three different directions in the search for a solution. The Navy focused on a "thin-film" circuit in which some components could be "printed" on a ceramic base, somewhat reducing the cost and size of the circuit; Jack Kilby worked on this idea for a while during his years in Milwaukee at Centralab. The Army's line of attack centered around the Micro-Module idea—the Lego block system in which different components could be snapped together to make any sort of circuit. Kilby worked on that one for a few days when he first arrived at Texas Instruments.

The Air Force, with a growing fleet of missiles that posed the most acute need for small but reliable electronics, came up with the most drastic strategy of all. It decided to jettison anything having to do with conventional circuits or conventional components and start over. The Air Force program was called "molecular electronics" because the scientists thought they could find something in the basic structure of the molecule that would serve the function of traditional resistors, diodes, etc. Bob Noyce brushed up against molecular electronics early in his career. "The idea of it was, well, you lay down a layer of this and a layer of that and maybe it will serve some function," Noyce said later. "It was absolutely the wrong way to solve anything. It wasn't built up from understandable elements. It didn't start with fundamentals because they were rejecting all the fundamentals. It was pretty clearly destined for failure." The Air Force wasn't listening. With strong lobbying from the generals, molecular electronics won the ultimate bureaucratic seal of approval—a line item of its own in the federal budget. Congress eventually appropriated some $5 million in research funds. Nothing came of the idea.

Each service, naturally, was eager to see its own approach prevail. All three services, consequently, were somewhat taken aback when they learned, in the fall of 1958, that a fellow named Kilby at Texas Instruments had worked up a solution to the numbers problem that was neither Army nor Navy nor Air Force.

The military services learned of Kilby's new monolithic circuit as soon as the people at Texas Instruments had tested the first chip and found that it worked. "TI had always followed a strategy of getting the Pentagon to help with development projects," Kilby explained later. "So sometime in the fall of 1958, Willis [Adcock] and I started telling the services what we had." The Navy wasn't interested. The Army agreed to provide funding, but only to prove that Kilby's new integrated circuit was "fully compatible" with the Micro-Module. "Well, it wasn't a Micro-Module at all," Kilby recalled. "But that was okay. It gave us some money to work with, and we didn't care what they called it. If they wanted it green, we'd paint it green." Hoping to supplement the modest Army grant, Adcock and Kilby spoke to the Air Force. "They weren't interested," Kilby said later. "Our circuit had the traditional components, resistors, and the like, and their approach wasn't going to have any of that traditional jazz." Despite the initial rejection, Adcock wouldn't give up. For months he argued his case, and eventually he found a colonel who was starting to lose faith in the cherished notion of molecular electronics. In June 1959 the Air Force agreed to help out, a little bit. Somewhat grudgingly, the service coughed up just over $1 million for developmental work on the chip, a piddling amount for a major new electronics project. (Years later, the Air Force's public relations wing put out a book on microelectronics: "The development of integrated circuits is, in large part, the story of imaginative and aggressive leadership by the U.S. Air Force.")

Events followed a different course at Fairchild, largely because Bob Noyce had different ideas about Pentagon-funded research. Noyce had worked on some defense research and development projects when he was a young engineer at Philco, and the experience left a sour taste that never went away. It wasn't fair, he thought—it was "almost an insult"—to ask a competent, creative engineer to work under the supervision of an Army officer who had at best a passing familiarity with electronics. The right way for the private sector to carry out research, Noyce felt strongly, was with private money. If this research happened to produce something useful for the military, fine, but Noyce did not want his engineers restricted to military research or bound by the confines of a defense development contract.

And so Fairchild developed the monolithic idea into a marketable commodity using its own funds. Noyce readily conceded, though, that the company was willing to do so in considerable part because of potential sales to the military market. "The missile program and the space race were heating up," Noyce said. "What that meant was there was a market for advanced devices at uneconomic prices . . . so there was a lot of motivation to produce this thing."

In addition to the Apollo program, several new families of nuclear missiles provided large early markets for integrated-circuit guidance computers. The designers of Minuteman II, the second-generation ICBM, decided in 1962 to switch to the chip. With that decision, which led to $24 million in electronics contracts over the next three years, the integrated circuit took off. Texas Instruments was soon selling 4,000 chips per month to the Minuteman program, and Fairchild, too, landed important Minuteman contracts. Soon thereafter the Navy began buying integrated circuits for its first submarine-launched intercontinental missile, the Polaris. By the mid-sixties, chips were routinely called for in specifications for a large variety of military electronic gear—not only G & N computers but also telemetry encoders, infrared trackers, loran receivers, avionics instruments, and much more. NASA's IMP satellite, launched late in 1963, was the first space vehicle to use integrated electronics, and thereafter chips became the circuits of choice in satellites and other space endeavors. About 500,000 integrated circuits were sold in 1963; sales quadrupled the next year, quadrupled again the year after that, and quadrupled again the year after that.

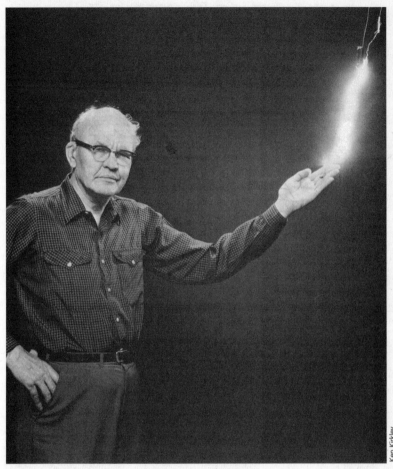

After ten years in the engineering boondocks, Jack Kilby (above), a quiet, slow-talking sort from Great Bend, Kansas, finally got a chance to work in a major laboratory on a problem of premier importance. Within weeks, he hit on an idea that struck the world of microelectronics like a lightning bolt.

In order to further test the feasibility of circuits on a single crystal wafer, I have had three units made up as shown:

These units were made from different material which had been prepared for development work on the 2N623. The wafers were lapped and the slot was etched using a special mask. Gold plated platinum tabs were alloyed to the back of the wafer. The emitter & base contacts, including those which serve as contacts for the capacitors, were evaporated on before the slot was etched. Platinum for the transistor and capacitor areas were masked, using wax, and etched. The bar was then cemented to a piece of microscope slide using Sauereisen cement. Gold wire was thermobonded to the contact areas, and the other end of the wire thermobonded to the gold tabs. These units were fabricated by John Yeargan from my sketch.

On testing, it was found that one unit, using a 7 volt supply, would act as a bi-stable device! That is by grounding the base lead alternately,

"It looks pretty crude," Kilby said when he looked back at the first integrated circuit. "In fact, it was pretty crude." But on September 12, 1958, Kilby was able to record in his lab notebook that this crude circuit-on-a-chip had worked perfectly on its first test.

Texas Instruments, Inc.

By the time Kilby received "Letters Patent" for his "Miniaturized Electronic Circuits," the right to the invention was already entangled in a labyrinthine legal snarl. The case finally turned on the infamous "flying wire picture." And the winner was . . .

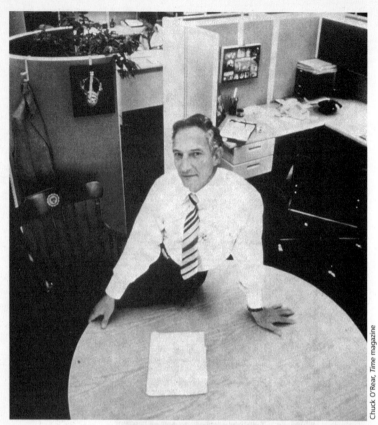

Chuck O'Rear, *Time magazine*

... Bob Noyce, the minister's son from Denmark, Iowa, a brilliant, multifaceted physicist and manager who hit on the monolithic idea independently. Noyce's first integrated circuit was based on the "planar" process, which eliminated the flying wires and made the chip a practical reality.

Fairchild Camera & Instrument Corp.

AT&T Bell Laboratories

Kilby and Noyce were spiritual descendants of another American genius, Thomas A. Edison, who first noticed the phenomenon of thermionic emission and built the "Edison Effect" lamp—a light bulb with a long glass nose sticking out—to study it.

It was the estimable J. J. Thomson (left), successor to Clerk Maxwell's chair at Cambridge, who finally explained what the Edison Effect was all about. In the process, J.J. discovered the electron and opened the door to a whole new scientific cosmos known as subatomic physics.

AT&T Bell Laboratories

National Park Service

A half-century after J.J.'s discovery, William Shockley (seated), John Bardeen (standing, left), and Walter Brattain eliminated the glass bulb and put electrons to work in the seminal invention of the semiconductor age: the transistor.

Patrick Haggerty (top, left) had the revolutionary notion that he could put the transistor into every home—by building the "transistor radio." The big radio makers scoffed at this harebrained idea, but Haggerty persuaded Edward Tutor (top, right), of tiny Regency Radio, to give it a try.

Ten years later, Haggerty decided he could put the microchip into every household as well. He told Jack Kilby to invent a portable adding machine. The result was the first pocket calculator (in the model's right hand, above). Today's little calculators make Kilby's breakthrough look like a dinosaur.

The first digital computers, like the University of Pennsylvania's ENIAC (above), filled large rooms and consumed the power of a locomotive to heat their thousands of vacuum tubes. Four decades later, a far more powerful computer (below) could fit on a user's lap.

The Microelectronics Revolution: from tube (left) to transistor (center) to microchip (right).

The burgeoning government sales not only provided profits for the chip makers but also conferred respectability. "From a marketing standpoint, Apollo and the Minuteman were ideal customers," Kilby said. "When they decided that they could use these solid circuits, that had quite an impact on a lot of people who bought electronic equipment. Both of those projects were recognized as outstanding engineering operations, and if the integrated circuit was good enough for them, well, that meant it was good enough for a lot of other people."

One of the major pastimes among professional economists is an apparently endless debate as to whether military-funded research helps or hurts the civilian economy. As a general matter, there seem to be enough arguments on both sides to keep the debaters fruitfully occupied for years to come. In the specific case of the integrated circuit, however, there is no doubt that the Pentagon's money produced real benefits for the civilian electronics business—and for civilian consumers. Unlike armored personnel carriers or nuclear cannon or zero-gravity food tubes, the electronic logic gates, radios, etc., that space and military programs use are fairly easily converted to earthbound civilian applications. The first chip sold for the commercial market—used in a Zenith hearing aid that went on sale in 1964—was the same integrated amplifier circuit used in the IMP satellite. For the Minuteman II missile, Texas Instruments had to design and produce twenty-two fairly standard types of circuits in integrated form; every one of those chips was readily adaptable to civilian computers, radio transmitters, and the like. A large number of the most familiar products of the microelectronic revolution, from the busy businessman's pocket beeper to the Action News Minicam ("film at eleven"), resulted directly from space and military development contracts.

The government's willingness to buy chips in quantity at premium prices provided the money the semiconductor firms

needed to hone their skills in designing and producing monolithic circuits. With their earnings from defense and space sales,
Fairchild, Texas Instruments, and a rapidly growing list of other
companies developed elaborate manufacturing facilities and precise new techniques for making chips. As experience taught ways
to solve the most common production problems, the cost of making a chip began to fall. By 1964 the initial manufacturing base
was in place, and the integrated circuit started flying down the
learning curve with the speed of a lunar rocket in reentry. In 1963
the price of an average chip was about $32. A year later the aver-
age price was $18.50, a year after that $8.33. By 1971, the tenth
anniversary of the chip's arrival in the marketplace, the average
price was $1.27. By the year 2000, a chip with the capacity of
those 1971 models would sell for a nickel or less.

While prices were falling, capability soared. Year after year,
buyers of integrated circuits got more product for less money.
Manufacturers learned how to cram more and more components
onto a single chip. This achievement was partly a matter of design; complex circuits had to be laid out on the tiny flake of silicon so that each individual component could perform its
function without interfering with any of the other components
squeezed alongside. The chief technical obstacle to high-density
chips, though, was production yield. The more components
printed on a chip, the greater the chance that one of those components would have a defect. One defective transistor could ren-
der the entire integrated circuit worthless. "A single speck of dust
is huge compared to the components in a high-density circuit,"
Bob Noyce said. "One dust particle will easily kill a whole circuit.
So you've got to produce the thing in a room that is absolutely free
of dust. You've got to build in thousands of [connecting] leads that
are finer than a human hair, and every one of them has to be free
of any defect. Well, how do you build a room that's free of dust?
And how do you print a lead that is essentially perfect? We had to

learn over time how to do things like that." Over time, the industry developed the "negative pressure" fabrication room—with a steady suction taking air, and dust, out of the room. The white nylon "bunny suit" that fab workers wear to prevent contamination has become a symbol of the microchip industry. And the machinery that "prints" circuitry onto CD-size "wafers" of silicon is so complex and so precise that a single photolithography unit costs tens of millions of dollars.

But as the industry learned how to operate at ever-tinier dimensions, it found itself in a delightful position. A chip containing 10,000 components required no more silicon, and not much more labor, than one with only 5,000 components. It was as if a fast-food stand had found a way to turn out two burgers using the same amount of meat and bread that it had previously used for one. The semiconductor industry found ways to double capacity over and over again. Noyce's friend and colleague Gordon Moore was asked in 1964, when the most advanced chips contained about 60 components, to predict how far the industry would advance in the next decade. "I did it sort of tongue-in-cheek," Moore recalled later. "I just noticed that the number of transistors on a chip had doubled for each of the last three years, so I said that rate would continue." To his dismay, that off-the-cuff prediction was widely quoted and soon came to be known as Moore's Law. This industrial "law" has developed all sorts of variants through the decades, but in its most common form, Moore's Law holds that the number of transistors on the most advanced integrated circuit will double every eighteen months or so. To Moore's astonishment, the law has held true all the way to the twenty-first century.

"At the time I said it, I had no idea that anybody would expect us to keep doubling [capacity] for even ten more years," Moore explained nearly four decades later. "If you extrapolated out ten years from then, to 1975, that would mean we'd have 65,000 transistors on a single integrated circuit. It just seemed ridiculous." By

1975 the industry was producing a new series of memory chips that contained 65,536 transistors. It takes roughly four discrete transistors to store one bit of information, so the 65,000-transistor memory chip stored 16,000 bits. This 16K chip gave way about three years later to the 64K memory chip, with some 258,000 transistors. With only a few hiccups along the way, progress in placing more and more components on a chip has closely followed Moore's Law ever since. By the first years of the twenty-first century, the state of the art in memory was a 256M chip, which held 256 million bits of information. The 256M circuit contains just over 1 billion transistors, all stacked on a sliver of silicon no bigger than the word "CHIP" on this page.

Memory chips tend to be the most crowded of all integrated circuits, because the transistors on those chips can be laid out in neat arrays of identical components. Other types of circuits are a little less heavily populated because their design requires less compact layouts. But these logic chips, too, have roughly followed Moore's Law, with the number of components doubling at roughly eighteen-month intervals. The Intel 8080 microprocessor that was coming to market about the time Gordon Moore made his prediction contained about 4,000 components. A quarter century later, the 8080's great-great-great-great-grandson, known as Pentium IV, had ten thousand times as many components— something over 40 million transistors—on a chip roughly the same size.

Around the time Jack Kilby won the Nobel Prize for the microchip, Gordon Moore took a long look back at the history of the invention and admitted that he is still amazed his own prediction continues to hold. "I still have a tough time believing that we can make these things," he said. "I'm a person who has been there literally since the beginning, and I know as a technical matter that a billion transistors on a chip is doable. Hell, we're doing it! But it is still astonishing we have come this far."

Moore's Law

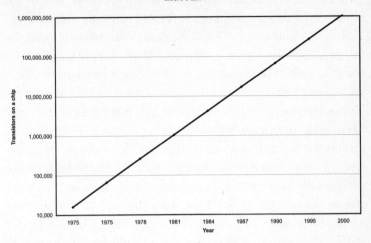

As fabrication plants turn out millions of chips that contain tens or hundreds of millions of transistors each, Moore—retired from active service at Intel, but still on the board—has spent some of his retirement time calculating just how many transistors are produced in a year. The comparisons he came up with demonstrate mainly how hard it is to imagine the numbers involved. At one point, Moore estimated that the number of transistors was greater than the number of raindrops that fall on California in an average year. Another calculation concluded that the semiconductor industry makes ten transistors per year for every ant on the earth.

With the steady decline in prices and the steep ascent in capacity since the birth of the microchip, the semiconductor industry has produced the greatest productivity gains in American industrial history. A graph comparing prices and capacity during the first forty years of the chip's existence makes a nearly perfect

X: the price curve angles sharply downward over time, and the capacity curve angles straight up. In the first generation of "solid circuits" back in the early 1950s, the chips were so simple and the prices so high that buyers were paying about $10 per transistor. By the year 2000, $10 would buy two 64-million-bit memory chips, with about half a million transistors. Clearly a 500,000-fold reduction in price is something special, and consequently it is probably unfair to compare the chip to other industrial products. The temptation is hard to resist, though, and the comparison is frequently made. In a typical version, Gordon Moore suggested what would have happened if the automobile industry had matched the semiconductor business for productivity. "We would cruise comfortably in our cars at 100,000 mph, getting 50,000 miles per gallon of gasoline," Moore said. "We would find it cheaper to throw away our Rolls-Royce and replace it than to park it downtown for the evening. . . . We could pass it down through several generations without any requirement for repair."

Another interested party who found this history hard to believe was Bob Noyce. "Progress has been astonishing, even to those of us who have been intimately engaged in the evolving technology," he wrote.

An individual integrated circuit on a chip perhaps a quarter of an inch square now can embrace more electronic elements than the most complex piece of electronic equipment that could be built in 1950. Today's microcomputer [on a chip], at a cost of $300, has more computing capacity than the first large electronic computer, ENIAC. It is 20 times faster, has a larger memory, is thousands of times more reliable, consumes the power of a light bulb rather than that of a locomotive, occupies 1/30,000 the volume and costs 1/10,000 as much. It is available by mail order or at your local hobby shop.

Noyce wrote that in 1977. Naturally, the passage is seriously out of date now. Today's microcomputer on a chip is vastly more powerful than the model Noyce had in mind, and costs considerably less.

The dramatic increase in the capacity of a chip also improved circuit performance. Higher-density chips meant less space between components. Less space meant less travel time for signal pulses running from one component to the next, so a smaller circuit was a faster circuit. Similarly, a smaller circuit required less power. Thus the chip makers were offering lower prices, higher capacity, and better performance year after year. In the second half of the 1960s that achievement caught the attention of commercial markets, such as producers of computers and industrial equipment, which had spurned the chip when it first appeared in 1961. By the early 1970s, the government was no longer the leading consumer of chips; that position had been taken over by the computer industry.

In retrospect, it is easy to see that the integrated circuit was perfectly suited to the digital computer, but the point was less than obvious to computer manufacturers when the chip first came on the market. In the days before the mini- and microcomputer, when computers routinely cost hundreds of thousands of dollars, the development of a new model represented an enormous investment on the manufacturer's part. Lead times were long; a decision made in 1961 would govern the production of machines that came on the market four years later. As a result, computer design was a conservative science. In the early 1960s no computer builder was willing to take a chance on a completely new type of circuit.

Thus when IBM brought out a major new line of computers, the System 360, in 1964, the new machine did not use integrated circuits. Still, the System 360 involved a revolutionary concept— it was a family of computers in various sizes and prices that shared the same instruction code, or software, and could thus communicate with one another—which rendered all existing machines obsolete. To fight back, competitors had to come up with something new of their own. They turned to integrated logic circuits. Using the chip, Univac, Burroughs, and RCA turned out machines that were as powerful as the IBM systems but smaller, faster, and cheaper. Brash upstarts like Digital Equipment and Data General entered the market with a new concept—the fully integrated device called the minicomputer. It was about the size of a senior executive's desk and cost less than $100,000 but matched some of IBM's big mainframes in computing power. In 1969, IBM bowed to the inevitable and began using chips for all logic circuitry in its computers. Now the chip makers had a market that would dwarf the space and defense business. By 1970, there were more than two dozen American firms turning out integrated circuits; they sold 300 million chips that year. Two years later 600 million chips were sold.

Logic circuitry, however, represented only one part of the potential computer market. Digital machines need logic gates to manipulate data, but they also need memory units to hold the data. The computers of the 1960s stored data, in the form of binary digits, using an ingeniously simple technique called magnetic core memory. A core memory looks like a tennis net made of fine wires; wherever two wires cross, a small iron wedding ring—the core—is hooked over the intersection. By sending electronic pulses along the right pair of wires, each individual iron core could be magnetized or demagnetized. A magnetized core represented a binary 1; demagnetized, it stood for binary 0. Core memory was fairly bulky; it took about 10 square feet of wire net to store 1,000 bits of information ("bit" is Claude Shannon's term for a single binary digit, a 1 or a 0). But it was reliable, easy to make, and inexpensive. The wires and the iron cores were dirt cheap, and a complete memory unit needed only a handful of transistors to send out the needed pulses. The most expensive thing about core memory was the labor cost for stringing all those iron rings on the net. This job, done by hand, was eventually farmed out to places like Hong Kong and Mexico, so prices for core memory remained low.

Some farsighted semiconductor engineers could see the possibility of putting memory onto a chip. The integrated circuit, after all, was a perfect medium for storing binary digits. A chip comprised a large array of switches (transistors), and any switch has memory. The light switch on the wall is a memory unit; it remembers the last thing you did to it, and stays that way—either on or off—until you change the setting. For various technical reasons, semiconductor memories often used more than one transistor to store each binary digit. In one standard memory design, a block of four transistors was used to store each bit. If a signal pulse turned the block on, it stood for 1; if the block of transistors was off, it stood for 0.

A semiconductor memory chip was ten times smaller than the equivalent core memory unit; since the signal pulses had shorter distances to travel, it was much faster. Through most of the 1960s, however, it was also two or three times more expensive. In 1967 engineers at Fairchild performed the prodigious feat of squeezing 1,024 transistors onto a single integrated circuit. At four transistors per bit, such a circuit could provide storage for 256 bits of information. But a 256-bit memory chip was still more than twice as expensive as a comparable amount of iron core memory. The Fairchild chip was admired in the laboratories but ignored in the market.

A monograph that appeared in the *Proceedings of the Institute of Electrical and Electronic Engineers* in 1968 set forth in discouraging detail the economics of the memory business. Just to approach the price of core memory, it said, the semiconductor people would have to come up with a 1,000-bit memory chip.

Semiconductor memory would not actually become cheaper until somebody developed a 4,000-bit chip. Four thousand bits on a single chip? Most of the industry looked at those figures and decided that the wisest course would be to forget memory. A pair of engineers at Fairchild—Bob Noyce and his friend Gordon Moore—looked at the same numbers and decided to give it a try.

By 1968, the men who had formed Fairchild Semiconductor were chafing at the controls imposed by their corporate superiors back east at the Fairchild Camera and Instrument Corporation. Noyce, Moore, and their colleagues knew more about the semiconductor industry, both the technical and the marketing side, than any of the Fairchild directors, but they were constantly forced to follow corporate decrees that seemed downright foolish. During 1967, moreover, the parent company went through a period of turmoil that saw two CEOs hired and fired within six months. To the "Fairchildren" out in California, it seemed obvious that the right man to lead the corporation was their own leader, Bob Noyce. But when this suggestion was passed to the corporate board, the directors could not bring themselves to entrust their established and traditional corporation to a California cowboy. When Noyce was passed over, the Californians gathered and agreed it was time to move on.

When Noyce, Moore, Andrew Grove, and several others left Fairchild in 1968 to start a new company specializing in semiconductor memories, they were gambling that a memory chip would be easier to make in extremely high densities than the traditional logic chip. A logic circuit, with its assorted gates and pathways, requires a variety of components laid out in complex patterns and a byzantine pattern of leads to connect the parts. A memory chip, in contrast, consisted for the most part of identical transistors lined up in identical blocks, one after another, like blocks of identical tract houses in suburban Levittown. Connections could be provided by a simple network of parallel leads, just like a neat grid of crisscrossing suburban streets. Each block of transistors, like each house in Levittown, could be assigned a unique address. Consequently, the memory circuit would permit random access — that is, the logic circuits could send data to or extract it from any one of 1,000 memory locations without disturbing the other 999.

The new firm that Noyce, Moore, and the others founded, Intel Corporation, turned out its first high-density memory circuit late in 1968. It held 1,024 bits of data. Using the standard engineering shorthand for 1,000, the letter "K," this random-access memory chip was called a 1K RAM. Intel rang up a grand total of $2,672 in sales that first year. By 1973, when the 4K RAM came to market, the firm's sales topped $60 million, and Texas Instruments and several other firms had jumped into the memory business as well.

Like most other fields of human endeavor, the computer world has its own version of Parkinson's Law. It is sometimes stated in pure Parkinsonian terms — "Data expands to fill the memory available to hold it" — and sometimes in plainer language — "There's no such thing as enough memory." For computer buffs, using one bank of memory is like eating one peanut. The computer business turned out to have a voracious appetite for cheap, fast random-access memory, and the semiconductor business geared up to meet the need. Partly because of the commonly used four-transistor-per-bit storage configuration, memory chips tended to grow by factors of four. A 16K RAM came on the market in 1975. The 64K RAM went on sale five years later. In accordance with Moore's Law, this growth has continued up to a 256M RAM chip at the start of the twenty-first century.

In addition to the two types of digital circuits — logic and memory — the late sixties also saw the first significant development of another species of chip, called a linear, or analog, integrated circuit. The linear chips replicated the functions of many traditional electronic circuits—timers, radio transmitters, audio amplifiers, and the like. Such applications put the integrated circuit into a number of noncomputer electronic devices. Some were traditional—the first integrated circuit radio receiver went on the market in 1966—and some wholly new—the cardiac pacemaker, a tiny circuit that gives off small electric pulses at precise intervals, was implanted in a human chest for the first time in 1967. Eventually, integrated electronics were replacing traditional circuitry in everything from elevators to Osterizers.

Applications for integrated circuits were multiplying rapidly, but not as rapidly as the semiconductor companies were turning out densely integrated new circuits. "We reached a point where we could produce more complexity than we could use," Gordon Moore said later. Those industry executives who took time to look up from their balance sheets and think about the future could see that supply of electronic circuits was increasing faster than demand. Almost alone among industrial products, moreover, the integrated circuit was a one-time-only sale. There was nothing to break, no moving parts to wear out. Once a chip passed its initial inspection, it would last a lifetime; there was little or no replacement market.

An exam question that pops up now and then in the nation's business schools postulates a situation where somebody invents a common product that will never wear out. The famous prototype is the miracle fabric that Alec Guinness invents in the great film *The Man in the White Suit*—a cloth that never gets dirty, wrinkles, or wears out. In the movie, the new product is at first welcomed with joy; then the dry cleaners, suit makers, and department stores of the world figure out what this breakthrough would do to them. In the end, the miracle fabric is quietly buried. But what if such a product came along in real life—a lifetime light bulb, for example, or permanent razor blade? How should the razor blade industry react? One answer is that the manufacturers should make a quick killing selling the lifetime blade until all conventional blades are replaced—and then go out of business. This answer is not an acceptable one, however, at most business schools, which preach the need for unending growth. The right answer is that the industry should use its ingenuity to create new uses for lifetime razor blades, and cater to a continually expanding market.

Faced with an ever-expanding supply of a lifetime product, the semiconductor industry at the end of the 1960s picked the right answer. The thing to do was to find new applications and new markets for integrated circuits. Since the chip, up to then, had been sold almost exclusively for government and industrial uses, the obvious new market to shoot for was the largest market of all — the consumer. To maintain its explosive rate of growth, the American semiconductor industry would have to take its revolutionary new product into the American home.

But how? The only chips that most Americans knew much about were made from potatoes; the very word "semiconductor" was completely alien to the general public. How many homes really needed an interplanetary guidance and navigation system or a $50,000 computer? Extending the microelectronic revolution down to the average consumer loomed as a formidable problem. To solve it, the industry turned to one of its premier problem solvers—Jack Kilby.

8

THE IMPLOSION

The decision that brought the chip into every household, and made "chip," in its microelectronic sense, a household word, was a carbon copy of the decision a decade earlier that had done the same things for the chip's immediate ancestor, the transistor. The decision maker in both cases was Patrick Haggerty, the farsighted and plucky chief executive of Texas Instruments. In the early 1950s, when the transistor was starting to become a cheap, reliable mass-production item, Haggerty developed a fascination, almost an obsession, with the notion that microelectronics should play a pervasive role in modern society. He came to believe that microelectronic devices would replace standard circuitry in existing electronic gear; that all tools and appliances controlled by traditional gears, springs, or timers would be fitted with chips; and that the availability of low-cost, low-power, high-reliability miniature components would create entire markets unknown before. "Pervasiveness" occurs like an idée fixe in Haggerty's speeches and writings. The conviction that microelectronic devices would gradually pervade every aspect of life was at the root of many of his business decisions.

Haggerty first put the principle into practice in 1954, when Texas Instruments, then a small regional manufacturer of electronic parts, discovered that it was ahead of everybody else in transistor development. "We knew we were doing pretty well in our semiconductor endeavor," Haggerty recalled twenty-five years later, "[but] we were facing a world that was pretty skeptical. . . . It seemed to me that it was imperative for T.I. to generate some kind of dramatic demonstration that reliable transistors really were available in mass production quantities and at moderate prices, and that T.I. was both ready and able to produce them." In fact, Haggerty had a particularly dramatic demonstration project in mind.

He had called in his engineers and told them to produce something wholly new—a portable radio, completely transistorized, powered by penlight batteries, and small enough to carry in a pocket or install in a dashboard. Haggerty tried to persuade several major radio firms to sell the device. They all demurred, arguing that there was no market for a pocket radio—an accurate assertion, since there had never been a pocket radio. A "radio" in 1954 tended to be a big wooden tabletop thing; it wouldn't fit in a suitcase, much less a pocket. But Haggerty persisted, and eventually found a small company, Regency, which introduced the pocket radio just in time for the Christmas sales rush in 1954. More than 100,000 Regency portables were sold the first year, and pocket radio sales increased astronomically thereafter. The product was as successful for TI as for Regency, because it made the transistor a familiar object throughout the world. Some Texas Instruments people say Haggerty's multimillion-dollar crash program to put the transistor into a consumer product was actually aimed at a single consumer—Thomas Watson Jr., the head of IBM. If that was Haggerty's aim, he hit a bull's-eye. Watson bought a hundred Regency radios and distributed them among his engineers; according to an IBM executive, Watson told his people that "if that little outfit down in Texas can make these radios work . . . they can make transistors that will make our computers work, too." Haggerty himself couldn't have put it better. In 1957, Watson signed a purchase order that made Texas Instruments a key IBM supplier and provided a huge new market for TI's transistors.

Eleven years after he had launched the radio, Haggerty got to thinking about the future of the integrated circuit. The chip was winning itself a niche in military and industrial markets but had yet to crack the computer industry or make even the smallest dent in the consumer market. Haggerty knew, and his engineers knew, that the chip represented a revolutionary advance in technology and that the product's reliability, capacity, and value were increasing every year—but the world at large did not know that. What was needed, Haggerty decided, was a dramatic demonstration of the benefits that monolithic circuitry could provide. And he had a particularly dramatic demonstration project in mind.

On an airplane trip with one of his engineers in the fall of 1965, Haggerty started talking about his belief that tiny, inexpensive, but complex circuits in integrated form would lead to new dimensions of pervasiveness. Why, the day would come, he said, when the chip would be built into a wide range of consumer products—when there would be integrated circuits in every home. As a matter of fact, he had been thinking about a consumer product that would be a perfect vehicle for the monolithic circuit. Before the plane landed, Haggerty had ordered the engineer to invent something wholly new: a miniature calculator that would fit in the palm of a hand—much lighter, much smaller, and much cheaper than any calculating machine that had ever been thought of before.

The man on the receiving end of that order was one of the most respected and fastest-rising engineers in the company—Jack Kilby. Since his invention of the integrated circuit during his first month at Texas Instruments, Kilby had received a series of raises and promotions befitting an employee who had given the company a firm position on the leading edge of monolithic technology. In a company where titles really mattered, Kilby had progressed from simply "Engineer" to "Manager of Engineering" to "Manager, Integrated Circuits Development" to "Deputy Director, Semiconductor Research and Development Laboratory." Best of all, from Kilby's point of view, the company had essentially left him alone to do the kind of work he liked best—solving technical problems. When Haggerty came up with the formidable problem of building a pocket calculator, Jack Kilby was the obvious person to take it on.

"I sort of defined Haggerty's goal to mean something that would fit in a coat pocket and sell for less than $100," Kilby recalled. "It would have to add, subtract, divide, and multiply, and maybe do a square root, it would have to use some sort of battery as a power supply, to make it portable, and it couldn't be too much heavier than a fairly small paperback book." It was a tall order. Today, of course, the calculator Kilby described is utterly commonplace, but in 1965 it was something quite unprecedented. There were calculating machines available then, but none came close to meeting Haggerty's terms. The standard electronic desk calculator was as large and as heavy as a full-size office typewriter. It contained racks and racks of electronic parts and dozens of feet of wire. It ran on 120 volts of electricity and cost roughly $1,200—about half the price of a family car.

The president's own brainchild naturally became a matter of some priority at Texas Instruments. The company, which yields nothing to the CIA when it comes to secrecy, put Kilby in a shrouded office and told him always to refer to his new project by a code name. An earlier TI research program had been called Project MIT, so Kilby took the logical next step and named his effort Project Cal Tech. "It was a miserable choice," Kilby recalled afterward. "Anybody who heard it would have figured out that we had a crash project going on calculator technology." In any case, Cal Tech soon grew into a team effort. Kilby started looking around the semiconductor lab for engineers who would not be daunted by the sizable technical problems involved in inventing a new species of calculator. Among others, he settled on a friendly, easygoing young Texan named Jerry Merryman.

Merryman, who had come to Texas Instruments two years earlier at the age of thirty-one, represents a vanishing breed in the high-tech industry—the self-taught engineer. After finishing high school in the country town of Hearne, Texas, he floated around and through Texas A&M for a few years but never stayed in place long enough to establish a major, much less earn a degree. Instead, he learned electrical engineering on odd jobs here and there and developed an almost intuitive sense for circuitry. "He's one of these guys who looks inside for a minute or two and then says, 'Well, maybe if we put that green wire over there,'" Kilby said. Like Kilby, Merryman has a fundamental confidence that any technical problem can be solved. "I just know," he said, "that you're going to find an answer if you think about it right. Eventually it'll come. A lot of inventions just happen on the way to work."

It was almost an act of faith in 1965 to believe that you could reduce the size, weight, and cost of an electronic calculator by factors of ten or more. One of the first things Kilby realized was that tearing apart existing adding machines to see how they worked—a process known as reverse engineering—would offer little, if any, help, because the basic architecture of this pocket-size device would have to be completely new. And so the team started at ground zero, setting down the fundamental elements that their calculator would require.

In accordance with the architecture worked out by Alan Turing and John von Neumann, all digital devices, from the most powerful mainframe supercomputer to the simplest handheld electronic game, can be divided into four basic parts serving four essential functions:

Input: The unit that receives information from a human operator, a sensory device, or another computer and delivers it to the processing unit. For the calculator, this meant a keyboard.

Memory: The unit that holds data—numbers, words, or instruction code—until the processing unit is ready to receive it.

Processor: The central control circuit that transfers data to and from various memory segments and manipulates the data. In a calculator, the processor performs the arithmetic.

Output: The display unit that shows the results of each calculation.

In addition to these four basic sections, Kilby had to worry about something that is no problem on nonportable electronic devices—a power supply. A calculator meant to be carried in a pocket and used anywhere could not be designed for the 120 volts of power available from a wall socket. Instead, the calculator would have to operate from a battery; it would be limited to about 5 volts.

Because Haggerty was in a hurry, and because Kilby had a basic confidence that the job could be done, the team decided on an all-points attack: the group would work on all their problems at once and hope that everything came together in the end. The input section—basically, the design of a small, power-efficient keyboard—was assigned to an engineer named James Van Tassel. Kilby himself took on the output section and the power supply. The memory and central processor—the calculator's electronic innards—were Merryman's responsibility.

"The basic rule was, everything had to be smaller than you'd

ever thought you could make it," Merryman said later. "Now, one
thing we did, we reduced the whole memory down to a single register. ["Register" is computerese for a short chain of transistors
that can store a dozen or so binary digits.] The only problem with
that was you'd need a lot of wires coming into that register, and
there wasn't much room inside that case for a lot of wires." To
solve that, Merryman designed a special "shift register"—a storage
circuit that is laid out something like a large auditorium with only
a single narrow aisle for people to come in and out. "That way, all
the bits could come in there, sort of march in single file, and since
they come in one at a time you get a lot of numbers in with just
one wire." Merryman's most serious problem, though, was with
the logic circuitry. The desktop electronic calculators of that time
employed thousands of logic gates—AND, OR, and NOT circuits—to carry out binary calculations. "We were trying to build
this whole circuit with only three bars," Merryman said ("bar" is
the engineer's slang term for a silicon chip). "That left me about
400 gates in all—maybe 4,000 transistors. And I worked out a processing unit that only needed 400 gates. It almost worked, too—
except it never could figure out how to get the decimal point in
the right place. That took a whole bunch of extra gates." In the
end, the team had to settle for a four-chip design with a total of
535 logic gates.

Meanwhile, Van Tassel had developed a working model for the
keyboard, and Kilby had found a rechargeable battery that would
run the device for three hours before running down. Memory,
processor, input, and power supply were in good shape. That left
only one problem—the output unit, for displaying the answer.
But this proved to be an unusually thorny problem.

Contemporary desk calculators used a cathode ray tube—a
miniature television set—for display, but such a system was far too
heavy, fragile, and power-hungry for a portable machine. For a
while, Kilby had hoped to use a row of tiny neon lights to display

the answers. As it turned out, that system required at least 40 volts—out of the question. Just down the hall from the Cal Tech team, a TI researcher was working on a new electronic device—a light-emitting diode, or LED—that was supposed to shine with a bright colored light when a minute current passed through it. This technology did, in fact, become the standard display technique for calculators and watches a few years later; in 1965, though, the diodes were not yet emitting much light.

There was nothing to do but invent something new. So Kilby developed a thermal printing technique, in which a low-power printing head would "burn" images into heat-sensitive paper; the idea worked perfectly, and the process is still widely used in low-cost, power-efficient printers.

All this activity consumed a little more than twelve months. One day late in 1966, Merryman recalled, "the thing was all laid out on the table like a person spread out on an operating table, all split open, wire running all over, and we punched in a problem, and it worked!" Silently, and almost instantaneously, the right answer came spinning out of the machine on a strip of paper tape. The Cal Tech group took their prototype in to Haggerty, who nodded with satisfaction—and called in the patent lawyers. It took another year before the design was perfected and the patent application—for a "Miniature Electronic Calculator"—could be filed. Although handheld calculators have come a long, long way since then, the Cal Tech team's architecture is still the gist of all such devices; even today many TI models carry the number of the original Kilby–Merryman–Van Tassel patent: 3,819,921.

The electronics of the new device were so far ahead of their time that it took years to turn out the initial production models. The world's first pocket calculator, the Pocketronic, was not introduced until April 1971—April 14, to be exact (the marketing people thought they might win the attention of taxpayers working late on Form 1040). By today's standards, that first model was a di-

nosaur—a four-function (add, subtract, multiply, divide) calculator that weighed 2½ pounds and cost about $150. But it sold like
crazy.

You would need a fairly high-powered calculator to keep track
of what happened next. Five million pocket calculators were sold
in the United States in 1972. As new features were added and
prices plummeted sales doubled year after year. To borrow a word
from Patrick Haggerty, the pocket calculator became pervasive.
Within a decade after the first pocket calculator was sold in the
United States, the country had more calculators than people. As
Haggerty had predicted, the new microelectronic gadget created
a market that had simply not existed before. Tens of millions of
people who never considered purchasing an adding machine or a
slide rule decided they wanted to own a pocket calculator. "How
many housewives actually need to know the square root of a number?" wrote Ernest Braun and Stuart MacDonald, two English
scholars who analyzed the phenomenon. "But then, the technology is ridiculously cheap. For a fraction of the cost of one week's
housekeeping, one can have permanent access to any number's
square root."

Today, with a four-function calculator—a model the industry
calls "plain vanilla"—available for $3.95 or so, the U.S. market is
virtually saturated. Yet Americans still buy between 26 million
and 30 million replacement calculators each year, including specialized models for stockbrokers, accountants, tax return prepar-
ers, bond traders, cattle ranchers, bicycle racers, cooks, and any
other market niche the salespeople can dream up. Worldwide, the
calculator market is a billion-dollar-per-year business, with sales
approaching 100 million calculators each year. In Japan, Casio
makes an abacus with a built-in calculator. Some people still use
the ancient counter to check the results that pop up on the calculator's display screen.

Another consumer application of the chip, born the same year as the handheld calculator and just as "pervasive" now, was the electronic, or digital, wristwatch. As a technical matter, the digital watch is markedly easier to make than a calculating machine. It is based on a convenient natural phenomenon called crystal oscillation, which is the physicist's way of saying that pure crystals of certain elements will oscillate, or vibrate back and forth, when connected to a source of electric current (e.g., a small battery). The rate of vibration depends on the atomic structure of the element; for a given material, though, the rate never varies. Certain crystals of the common mineral quartz, for example, will vibrate back and forth, back-and-forth, precisely 3,579,545 times each second.

A precise oscillator, be it the 5-foot brass pendulum of a grandfather clock or the .5-cm flake of quartz in a wristwatch, is the heart of any timepiece. All the watchmaker needs is a mechanism to count the back-and-forth oscillations—and counting is one of the simple tasks that binary logic gates can perform. In the digital watch a logic gate called a JK flip-flop counts the vibrations of the crystal. Every time the count hits 3,579,545, the gate sends a pulse to the display unit and the watch records the passage of another second. Another set of gates on the same chip counts 60 seconds and updates the minute display; another counts minutes to update the hour. If you tore apart your digital watch (why not? you can get a new one for five bucks), you would find, in place of the gears, springs, bearings, and bushings of a traditional timepiece, only four parts: a battery, a crystal, a chip, and the display unit.

Nonetheless, the first digital watch—produced by an American firm under the Pulsar brand name and introduced in the fall of 1971—was marketed as a decidedly high-bracket item. The 24-karat gold Pulsar was priced at $2,000; a stainless steel model cost $275. Characteristically, as the electronic watch improved—getting smaller, easier to read, more power-efficient—its price fell.

The American reader may have spotted in this history a perfectly legitimate excuse for chauvinism. Despite the predominance of names like Sony and Seiko, Canon and Casio, the major consumer products of the microelectronic age all resulted from pure Yankee ingenuity, as did the fundamental breakthrough—the monolithic idea—that made such advances possible in the first place. Polls show that many Americans consider the microelectronic revolution just another import from Japan—one more manifestation of the Japanese genius for technology and marketing. In fact, the flow of genius has gone in the other direction. The history of microelectronics has been a history of Japanese firms—and other companies around the world—learning at the feet of American innovators. This familiar pattern was played out once again with the development of the device that has taken microelectronics further than ever down the path of pervasiveness—the microprocessor.

The story of the microprocessor begins in Tokyo, but the scene shifts rapidly to Silicon Valley. In 1969 a Japanese business-machine manufacturer, Busicom, was planning a new family of desktop printer calculators but could find no engineers in Japan capable of designing the complex set of integrated circuits the machines would require. Busicom sought help—from Bob Noyce, who was still putting together his new company, Intel. The Japanese signed a contract with Intel calling for the design and production of twelve interlinked chips for the new line of machines. Busicom sent a team of engineers to Intel to oversee the work. Noyce, meanwhile, handed the problem to a one-man team—Marcian E. "Ted" Hoff, a thirty-four-year-old Ph.D. who had been lured away from a teaching job at Stanford by the prospect of broader horizons in industry. Although Hoff was an expert on microcircuits, his real ambitions were somewhat larger. He had always wanted to design his own computer.

When the Busicom engineers showed Hoff their tentative

plans for the twelve chips they needed, the American was appalled. The arrangement was outrageously complex—some of the simplest functions would require sending the same number into and out of two or three different memory registers—and could not possibly be implemented at an acceptable price. Even worse, in Hoff's eyes, the design was inelegant. It was downright wasteful to put dozens of man-years into designing a set of specialized circuits that could be used in only one small group of machines.

This last concern was important to Noyce as well. By the end of the sixties, Noyce was worried about the rapid proliferation of different integrated circuits, each designed for its own special purpose. Every customer who wanted a chip for his product was demanding a custom-designed chip just for that product. "If this continued," Noyce and Hoff wrote later, "the number of circuits needed would proliferate beyond the number of circuit designers. At the same time, the relative usage of each circuit would fall. . . . Increased design cost and diminished usage would prevent manufacturers from amortizing costs over a large user population and would cut off the advantages of the learning curve."

Looking ahead, Noyce saw that the solution to proliferation of special-purpose integrated circuits would be the development of general-purpose chips that could be manufactured in huge quantities and adapted ("programmed") for specific applications. Hoff had been intrigued by this concept and was frankly looking for an opportunity to give it a try. When the Busicom assignment landed in his lap, he grabbed the chance. Scrapping Busicom's ideas, the designer came up with a strikingly new design for the Japanese: a general-purpose processor circuit that could be programmed for a variety of jobs, including the performance of arithmetic in Busicom's machines. As Hoff pointed out, this approach would permit much simpler circuitry than the Japanese firm had suggested. Indeed, by the summer of 1971, Hoff was able to put all the logic circuitry of a calculator's central processor unit, or CPU, on a single chip. The CPU could be coupled with one chip for memory, one for storage registers, and one to hold the program; the entire family of calculators would require only four integrated circuits. In his job as a circuit designer, Hoff had, in fact, fulfilled his personal dream: he had designed his own general-purpose computer.

By the summer of 1971, though, the calculator industry was in the throes of great change. The introduction of Jack Kilby's $150 handheld calculator that spring had completely changed the rules; companies like Busicom with their heavyweight $1,000 machines were in big trouble. Accordingly, Busicom told Intel it could no longer pay the price originally agreed upon for the new chips. Negotiations ensued. Busicom got its lower price, but gave up something in return that turned out to be almost priceless: its exclusive right to the chips. Intel was now free to sell Hoff's general-purpose CPU-on-a-chip to anybody.

But would anybody buy it? That question spurred furious debate at Intel. The marketing people could see no value in a one-chip CPU. At best, a few minicomputer firms might buy a few thousand of the chips each year; that wouldn't even pay for the advertising. Some directors were worried that the new circuit was too far afield from Intel's real business. Intel, after all, was a circuit maker; Hoff's new chip was a single circuit, all right, but it really amounted to a complete system—almost a whole computer. There was strong pressure, Noyce and Hoff wrote later, to drop the whole thing.

Intel had recently hired a new marketing manager, Ed Gelbach, and he arrived at the company in the midst of this controversy. As it happened, Gelbach had started in the semiconductor business at Texas Instruments; like everyone else at TI, he was steeped in Patrick Haggerty's view of the world. Gelbach realized immediately that Intel had reversed the course of the industry by producing a general-purpose chip. "General purpose," Gelbach saw, was just another way of saying "pervasive." The real markets for the new device, he said, would be completely new markets. With this one-chip central processor—known today as a microprocessor—the integrated circuit could "insert intelligence into many products for the first time."

And so Intel's new "4004" integrated circuit went on sale for $200 late in 1971. With mild hyperbole, Intel advertised the device as a "computer on a chip." Gradually, as people realized that it really could work just about anywhere, the microprocessor started showing up just about everywhere. A typical application was the world's first "smart" traffic light. It could tell, through sound and light sensors, when rush hour was starting, peaking, or running down; the tiny CPU would alter the timing of red and green in response to conditions to maximize traffic flow. Soon there was a smart elevator, a smart butcher's scale, a smart gas pump, a smart cattle feeder, a smart intravenous needle, and a bewildering array of other "smart" devices. A microprocessor in a K2 ski would react to vibrations and stiffen the ski laterally, reducing bounce on the run. A microprocessor in a tennis racket senses where the ball has hit the racket and instantly adjusts string tension to make that very point the racket's "sweet spot" for that one shot.

Texas Instruments, of course, was hardly pleased to let its archrival steal a march on such an important new battleground. Working on a contract for a customer who wanted circuitry for a "smart" data terminal—a keyboard-screen combination that could communicate with large computers far away—a TI engineer named Gary Boone developed a slightly different version of a single-chip processor unit. Boone's version, called the TMS 1000, received the first patent awarded for a microprocessor. Then, at the end of 1971, Boone and another Texas Instruments engineer, Michael Cochran, produced the first prototype of an integrated circuit that actually was a computer on a chip. The single monolithic circuit contained all four basic parts of a computer: input circuits, memory, a central processor that could manipulate data, and output circuits. A year later, Intel came out with a second-generation microprocessor; since it had roughly twice the capacity of the original 4004 chip, the new device was called the 8008. The introductory price was $200. That morphed into the 8080, the 8086, and then a series of progressively more powerful processor chips that powered progressively more powerful personal computers: the 80286, 80386, 80486, Pentium, Pentium Pro, Pentium III, and Pentium IV. The Pentium IV chip operated at a speed about five hundred times faster than the 8008; the price was about the same.

It was the marriage of the microprocessor and a group of devices called transducers that finally brought microelectronics into every home, school, and business. A transducer is an energy translator; it converts one form of energy into another. A telephone receiver is a transducer, changing your voice into electrical pulses that travel through the wire. The keyboard on a calculator converts physical pressure from a finger into pulses that the central processor can understand. Other sensors can turn sound, heat, light, moisture, and chemical stimuli into electronic impulses. This information can be sent to a microprocessor that decides, according to preprogrammed directions, how to react to changes in its environment.

A heat-sensitive transducer can tell whether a car's engine is burning fuel at peak efficiency; if it is not, the transducer sends a pulse to logic gates in a microprocessor that adjust the carburetor to get the optimum mixture of fuel and air. A light-sensitive transducer—the familiar electric eye—at the checkout stand reads the Universal Product Code on a carton of milk and sends a stream of binary pulses to a microprocessor inside the cash register. The central processor queries memory to find out the price assigned to that specific product code today, adds that price to the total bill, and waits patiently (this has all taken three thousandths of a second) for the transducer to read the next product code. A moisture sensor and a heat sensor inside the clothes dryer constantly measure the wet clothes and adjust the machinery so that the laundry will be finished in the shortest possible time.

Microelectronics is at work inside the human body. A microprocessor that controls a speech-synthesizing chip can be connected to a palm-size keyboard that permits the mute to speak. Now under development is a chip that may be able to turn sound into impulses the brain can understand—not just an electronic hearing aid, but an electronic ear that can replace a faulty organic version. Other experiments suggest the possibility of an implantable seeing-eye chip for the blind—a light-sensitive transducer connected to a microprocessor that sends intelligible impulses to the brain.

Among the countless new applications that people dreamed up for the computer-on-a-chip was, of all things, a computer—a completely new computer designed, not for big corporations or mighty bureaucracies, but rather for ordinary people. The personal computer got its start in the January 1975 issue of *Popular Electronics* magazine, a journal widely read among ham radio buffs and electronics hobbyists. The cover of that issue trumpeted a "Project Breakthrough! World's First Minicomputer Kit to Rival Commercial Models." Inside, the reader found plans for a homemade "microcomputer" in which the Intel 8080 microprocessor replaced hundreds of individual logic chips found in the standard office computer of the day. The *Popular Electronics* kit was strictly bare-bones, but it gave anybody who was handy with a soldering iron the chance to have a computer—for a total investment of about $800. At a time when the smallest available commercial model sold for some $30,000, that was indeed a breakthrough. Readers sent in by the thousands. The other electronics magazines started offering computer kits of their own. Within a year thousands of Americans were tinkering with their own microprocessor-based personal computers.

The personal computer community in those pioneering days was a sort of national cooperative, with each new computer user eagerly sharing techniques and programs with everybody else. It was a community that lived by the old Leninist maxim, "From each according to his ability, to each according to his need." If you needed a program that would make your homebrew computer compute square roots, and if I had the ability to write a program to do just that, I would proudly share my handiwork with you—for free. But among the first to start programming the *Popular Electronics* 8080-based computer was a Harvard undergraduate who had a different idea. In the late 1970s he wrote the first genuinely useful program for 8080-based PCs, a simple version of the BASIC programming language. And then he did an amazing thing: he charged money for it. For this, the young programmer was attacked and vilified by many of his fellow buffs. And yet there were people willing to pay $50 for the BASIC program. Sales grew so fast that the undergraduate dropped out of college, to the despair of his parents, and started a tiny business selling software for the new breed of "microcomputers." Bill Gates named his company Microsoft.

These early computer buffs—"addicts" might be a more descriptive term—began forming clubs where they could get together for endless debates about the best approach to bit-mapped graphics or the proper interface for a floppy disk or the relative merits of the 8080, 6502, and Z80 microprocessors. At one such organization in Silicon Valley, a group called the Homebrew Computer Club, two young computer-philes, Steven Jobs and Stephen Wozniak, convinced themselves that there had to be a larger market for personal computers than the relatively small world of electronics tinkerers. The gimmick, they decided, was to design a machine that was pretty to look at and simple to use. The important thing was that the personal computer could not be intimidating; even the name of the machine would have to sound congenial. Eventually, Jobs settled on the friendliest word he could think of—"apple." It had nothing to do with computers or electronics, but then that was the whole point. The two started a computer company called Apple.

There was a time—when computers were huge, impossibly expensive, and daunting even to experts—when the sociological savants regularly warned that ordinary people could become pawns at the hands of the few corporate and governmental Big Brothers that could afford and understand computers. This centralization of power in the hands of the computer's controllers was a basic precept of Orwell's *1984*. But by the time the real 1984 rolled around, the mass distribution of microelectronics had spawned a massive decentralization of computing power. In the real 1984, millions of ordinary people could match the governmental or corporate computer bit for bit. In the real 1984, the stereotypical computer user had become a Little Brother seated at the keyboard to write his seventh-grade science report.

Patrick Haggerty, the visionary who had predicted that the chip would become "pervasive," had been proven right. By the twenty-first century, microelectronics did pervade nearly every aspect of society, replacing traditional means of control in familiar devices and creating new aspects of human activity that were previously unknown. By shrinking from the room-size ENIAC to the pinhead-size microprocessor, the computer had imploded into the basic fabric of daily life.

Haggerty lived until 1980, long enough to see his prediction starting to come true, but not to determine what the impact would be. His successors struggled to grapple with that issue.

DIM-I

A small but noteworthy segment of American industry, dead of competitive causes, was formally laid to rest in Washington, D.C., on a summer day in 1976. The rite of internment was a fittingly sad ceremony at the Smithsonian Institution. Keuffel & Esser Company, the venerable manufacturer of precision instruments, presented the museum with the last Keuffel & Esser slide rule, together with the milling machine the company had used to turn out millions of slide rules over the years for students, scientists, architects, and engineers. Shortly past its 300th birthday—the rule was invented by the seventeenth-century British scholar William Oughtred, whose other great contribution to mathematics was the first use of the symbol "×" for multiplication—the slide rule had become a martyr to microelectronic progress.

"Progress," in this case, meant Jack Kilby's handheld calculator. In the five years after the small electronic calculator first hit the market, K & E, the largest and most famous slide rule maker in the world, had watched sales fall from about 20,000 rules per month to barely 1,000 each year. Toward the end, one of the major sources of slide rule sales was nostalgia, as museums, collectors, and photographers bought up the relics on the theory that the slide rule would soon disappear. By the mid-seventies, the handwriting was on the screen, so to speak, and even Keuffel & Esser was selling its own brand of electronic calculator. "Calculator usage is now 100 percent here," an MIT professor told *The New York Times* in 1976, and that statement was essentially the obituary of the slide rule.

K & E's last slide rule was eventually deposited in a bin on a storage shelf at the Smithsonian's National Museum of American History. Someday it may be dusted off and put on display; at present, a special appointment must be made to view the relic. The curator says almost nobody ever bothers. Still, the slide rule lives on, in the affectionate memory (and frequently, amid the clutter in the desk drawers) of a whole generation of scientists and engineers, Jack Kilby among them. When word came in the fall of 2000 that Jack had won the Nobel Prize, the photographers who showed up at his office insisted that Kilby pose with his old slide rule, roughly the equivalent of asking Henry Ford to pose on horseback. Kilby and other engineers of his vintage recall the slide rule today with the same fond regard that an old golfer might have for a hickory-shafted mashie niblick or an auto buff reserves for the original 1964 Ford Mustang. In a requiem for the slide rule published in *Technology Review,* Professor Henry Petroski recalled that the Keuffel & Esser Log-Log Duplex Decitrig he bought as an undergraduate in the 1950s became his most valuable possession. "That silent computational partner [was] my constant companion throughout college and my early engineering career."

From a practical viewpoint, though, the competition between the slide rule and the calculator was completely one-sided from the beginning. The slide rule was essentially a complicated ruler. In the most common form, it was a ten-inch-long rectangle made of ivory, wood, or plastic with three different numerical scales marked along it—one on the top edge, a different on the bottom edge, and another in the middle. The middle section could slide right and left between the top and bottom. While a ruler is marked off in equal increments—1 inch, 2 inches, 3 inches, with each whole number exactly an inch apart—a slide rule was calibrated in the logarithms of the whole numbers. A "logarithm" is a tiny bit of mathematical magic. By sliding a piece of plastic marked in logarithms, you can multiply, divide, square, cube, or find the square root of any number. So the slide rule, for three centuries, served as a simple calculator.

The downside was that the slide rule gave only approximate answers; if you used it, for example, to calculate the square root of 470, it could tell you that the answer is somewhere around 21.6 or 21.7, but couldn't get much closer. And it offered no help at all in solving some of the trickier aspects of calculation—determining the order of magnitude and putting the decimal in the right place. The most primitive $3.99 calculator, in contrast, can solve problems precisely, down to the last decimal with a speed and accuracy that no slide rule could match. The square root of 470? Push two buttons and the answer leaps to the display screen: 21.67948338868. Today, in the hazy afterglow of memory, engineers tend to look on the slide rule's drawbacks and see virtues. "The absence of a decimal point," Professor Petroski wrote, "meant that the engineer always had to make a quick mental calculation independent of the calculating instrument to establish whether the job required 2.35, 23.5, or 235 yards of concrete. In this way, engineers learned early an intuitive appreciation of magnitudes. Now the decimal point floats across the display of an electronic device among extended digits that are too often copied down without thought." He has a point, of course; it's obviously a good idea for engineers or contractors to develop an "intuitive appreciation of magnitudes." On the other hand, this feature of the slide rule probably didn't seem so charming to a construction foreman who suddenly found himself with 200 excess tons of concrete hardening on his job site.

The slide rule was a simple instrument (at least, if you knew how logarithms work). To hear engineers tell it, that simplicity was part of its appeal. "It has a sort of honesty about it," Jack Kilby told me one day, reaching into the drawer of his desk and pulling out his old K & E. "With the slide rule, there're no hidden parts. There's no black box. There's nothing going on that isn't right there on the table." To put it another way, the slide rule was not threatening. Nobody ever called the slide rule a "mechanical brain." Nobody ever declared that the slide rule was endowed with something called artificial intelligence. There were no movies about runaway slide rules called HAL seizing control of the spaceship or plotting to dominate mankind. The slide rule, hanging at the ready from the belts of Fermi, Wigner, and Wernher von Braun, helped men create the first nuclear chain reaction and send rockets to the stratosphere. But the rule was always recognized as nothing more than a tool. It had no more "intelligence" than a yardstick or a screwdriver or any other familiar tool that extends human power. Like the yardstick, the screwdriver, etc., the slide rule was just an ignorant mechanism.

Someday—fairly soon, probably, given the accelerated pace of technological development—the pocket calculator, the handheld computer, the industrial robot, the cell phone, and other digital marvels of our day will themselves be museum pieces, on exhibit in a gallery called Primitive Microelectronic Tools or some such. As we file past with our grandchildren, we may well break into nostalgic smiles of fond regard for these devices that used to be considered so revolutionary. The kids, no doubt, will be amazed to learn that back at the turn of the twenty-first century many people still resisted those simple tools, resented them, feared them— feared that digital computers, robots, etc., and their so-called artificial intelligence might replace poor bungling man as the reigning intelligence on earth. To our grandchildren's generation, how foolish this will seem! Why, they will wonder, would anyone have feared such an ignorant mechanism?

To most of us today, even the simplest digital device seems incomprehensible. It is, as Jack Kilby has suggested, a black box. What goes on inside the black box is, for most people, black magic. You push the keys. The answer to some impossibly difficult math problem shows up on the screen, instantly. The rest is mystery. It's no wonder that calculators, computers, etc., are thought of as intelligent machines; what other explanation could there be? In fact, the explanation of how the computer gets the answer is not at all magical. The "magic" inside the black box actually involves a series of mathematical and logical techniques carried out by artful arrangements of electronic switches arrayed in logic gates. Electronic impulses are pulled this way and that through the maze of electronic switches by blind physical force.

The electrons racing through a computer chip have as much intelligence as water running down a hill. Gravity pulls the water. Electricity—the attraction and repulsion of electronic charges— pulls the electrons. If people build sluice gates and irrigation canals in the right combinations, they can make water flow where it is needed to water the fields. If people build logic gates and connecting leads in the right patterns, they can make electronic impulses flow where they're needed to solve a problem. In each case, the mechanism does the work, but in each case, it's an ignorant mechanism. The human designer provides all the intelligence.

If you punch into your calculator the task of adding 3 + 2, the mechanism will produce the answer 5. The calculator gets the answer not because of "artificial intelligence" but rather because a genuine intelligence—the human mind—has designed the mechanism so that it gets the right answer.

Using switching logic (from the minds of George Boole and Claude Shannon) implemented by transistors (from the minds of

Shockley, Bardeen, and Brattain) contained in the monolithic circuit (from the minds of Kilby and Noyce), the humans who build digital machines have designed an addition circuit in such a way that the pattern of pulses representing binary 5 is the only possible combination that can come out when binary 3 and binary 2 are put in. To get that sum, however, out of a mechanism consisting entirely of switches turning on and off, off and on, humans have had to go to some extreme lengths. For a machine as dim-witted as a computer to solve 3 + 2, the problem must be broken down into an absurdly detailed sequence of instructions that lead the machinery through its paces, step by elementary step. If there is magic in the pocket calculator, it is not in the machinery; it is in the humans who had the wit and the patience to program the machine to do its job.

To bring the point home, we can take a guided tour through the interior of the black box and watch a typical digital mechanism from the inside as it does its stuff. We'll look at a simple pocket calculator—so simple it exists only in the pages of this book—called the Digital Ignorant Mechanism, Model I, or DIM-I for short. The design of DIM-I to be set forth here is based on the familiar four-function calculator available anywhere for $5 or so. To a considerable extent, though, the basic architecture of a $5 calculator is the same as that in a $50 video game, a $500 handheld computer, a $5,000 corporate server, or the $5 million supercomputers that guide NASA's rockets through the cosmos. The bigger, more expensive machines can handle more information, store more results, and deal with a larger variety of tasks, but the modus operandi is the same. If you've seen one digital ignorant machine at work, you've pretty much seen them all.

On the outside, DIM-I is in fact a black plastic box. The box has eighteen keys: one each for the digits 0 through 9, and eight others for functions like +, =, etc. It has a display screen that can show numbers up to eight digits long. It's rather light—a few ounces at most—and if you were to pry open the black box, you'd see why. There's almost nothing inside. There's an empty space containing a tiny power source—a battery sometimes, or a solar energy converter—a few wires, and a printed circuit board on which sits another, smaller black box. This one is a piece of plastic that looks like a man-made millipede: an inch-long rectangle with a symmetrical array of wire legs sticking out from each side. That millipede is the chip—or, more precisely, the plastic package that holds the chip. The two rows of legs along the sides are the electronic leads that connect the keyboard and the display screen to the chip.

Inside a computer—even the smallest handheld computer—are whole platoons of these small black millipedes, each chip designed for a specific function. Open the back of a desktop computer, for example, and you can count about 200 separate chips. There are memory chips, logic chips, input-output chips, and microprocessor chips, all lined up in formation on the various circuit boards. Part of the miracle of microelectronics is that more and more separate functional elements can be squeezed into a single chip; a simple drugstore calculator like DIM-I uses just one chip that has all the necessary functional elements built into it.

Just for simplicity's sake, we'll say that the chip inside DIM-I is a TMS 1000C, one of the common microprocessors designed for small calculators. It has (1) a set of logic gates that read electronic signals from the keyboard and "encode" them in binary form that a calculator can understand. It has (2) a relatively small number of memory units—chains of transistors lined up in ordered rows so that each one can be addressed separately. That means the machine can randomly get to each memory block, so this is random-access memory, or RAM. It has (3) an arithmetic processor unit—a group of transistors arranged in gates so that they can use Boolean logic to perform simple math. And it has (4) a set of gates that "decode" binary information back into decimal form and send it to the display screen. In other words, the chip has the necessary circuitry to do four things:

1. Sense numbers punched into its keyboard.
2. Write them on an electronic scratch pad.
3. Add, subtract, divide, and multiply them.
4. Report the sum in the form of lighted digits on the display screen.

Which is just another way of saying that DIM-I can carry out the four essential jobs of any digital device:

1. input
2. memory
3. processing
4. output

A fancier machine—say, DIM-II or DIM-III—might be able to handle larger numbers, produce graphs on the screen, and store greater quantities of information in memory. By the time one gets to DIM-X or DIM-XLVII, the machine can deal with words as well as numbers and manipulate the information in countless ways that are beyond the wildest dreams of our little calculator. DIM-I, by contrast, can't do much. But it can demonstrate to us the basic computational mechanics, because what it does do it does in the same way as every other calculator and computer.

The processing circuitry of DIM-I, like that in any digital device, also has a central set of logic gates called the control unit. This is a sort of central switchboard that busily directs electronic pulses here and there, from input to memory to processor to the display screen, as needed to,solve the problem. The control unit is itself controlled, in turn, by a simple, familiar device that is essential to the operation of any digital machine—a clock.

To the extent that we attribute anthropomorphic characteristics at all to computers, the proper analogy would be not that the machine has a brain, but rather that it has a heart—a steady, pulsing central rhythm instrument that orchestrates and controls everything that happens. The computer engineers call the central clock a clock generator, because it really is a circuit that generates pulses at a perfectly steady, unvarying rate. The clock is to the computer what the bandleader is to the band; it stands there keeping time, ONE-two-three, ONE-two-three, so that each part will come in at the right beat.

The clock inside DIM-I beats, without variation, every ten millionth of a second—that is, it emits 100,000 pulses every second, or one pulse every 1/100,000 second. This is a fairly standard clock rate for cheap, simple pocket calculators. It is slow compared to the operating speed of large mainframe computers—or of small laptop computers, for that matter—but extremely fast compared to, say, a flash of lightning or the blink of an eye. An eye's blink takes about .30 second; in the blink of an eye, the clock inside a simple calculator will have pulsed 30,000 times.

There is no way for humans, in our poky world of seconds, minutes, and hours, to conceive of a time period like 1/100,000 second, much less the microsecond (1/1,000,000 second), the nanosecond (1/1,000,000,000 second), the picosecond (1/1,000,000,000,000 second), or the femtosecond (1/1,000,000,000,000,000 second). On the human scale, anything that lasts less than about a tenth of a second passes by too quickly for the brain to form a visual image, and is thus invisible; if the duration is less than a thousandth of a second or so, the event becomes too fast even for subliminal perception and is completely outside the human sphere. The speed of microelectronic events puts them in a world far removed from the human realm; how can an engineer contemplate a thousandth of a thousandth of a second? Computer engineers, practical types not often given to metaphysical speculation, don't even try. They just become, as Bob Noyce said, "reconciled" to the notion that their machines work at unthinkably high speeds.

The inconceivable speed of operation comes about because the "moving parts" of a digital machine are electronic pulses that travel inconceivably fast over distances inconceivably small. An electric signal moves at the speed of light, 186,000 miles per second. This makes for extraordinarily rapid transit, even at transcontinental distances.

A baseball fan in Boston turns on his TV to watch the big game in Los Angeles. He watches the pitch coming in, the hitter start to swing, and the smack of bat against ball. He does not see this at the precise instant it happens, though. The electronic signals carrying the picture to his television take about .016 second to travel the 3,000 miles from America's West Coast to the East. Thus the viewer will not see the bat hit the ball until about 1/62 second after the impact actually occurs. (For a fan watching the game in the form of streaming video over the Internet, the scene may be delayed even more—as much as half a second more—by the software that converts the impulse to a video image on the computer screen.)

The signals that turn switches on and off inside our calculator also move at the speed of light, but they do not have to travel 3,000 miles. On a quarter-inch square integrated circuit containing 10,000 components, individual transistors are spaced a few ten thousandths of an inch apart. Electronic charges moving at the speed of light traverse those distances in tiny periods of time. Consequently, the clock generator that controls the dispatch of pulses around the chip can be set to tick at tiny intervals.

In setting the clock rate for digital machines, however, the human designers have to consider not only the transit time for pulses racing from one transistor to another but also the time it takes for each transistor to switch from on to off. The computer is made of a long, long chain of transistors; each one has to wait for the transistor ahead of it in line to flip one way or the other before it can do anything. In most modern computers, the transistors' switching time (known to engineers as propagation delay) is a more serious constraint than the travel time for signals moving through the circuit.

"Propagation delay," like many other elements of microelectronic jargon, is a complex term for a simple and familiar phenomenon. Propagation delays occur on the freeway every day at rush hour. If 500 cars are proceeding bumper to bumper and the first car stops, all the others stop as well. When the first driver puts his foot back on the accelerator, the driver in the second car sees the brake lights ahead of her go off, and she, in turn, switches her foot from brake to accelerator. This switching action, from brake to accelerator, is then relayed down the chain of cars. If each driver has a switching time of just one second, the last car will have to wait 500 seconds—about 8½ minutes—because of the propagation delay down the line of traffic.

The transistors inside a digital device go from brake to accelerator—from off to on—somewhat faster. In the most high-powered machines, the switching time is about one nanosecond—one billionth of a second. In the current generation of personal computers, using the Pentium IV microprocessor, the transistors switch in about two billionths of a second. By those standards, a cheap pocket calculator like DIM-I is a tortoise. Its transistors take about 5 microseconds to go from on to off. To account for that much propagation delay and to allow a little additional time for signals to travel through the circuitry on the chip, the clock rate—the time signature that sets the tempo for each operation—in a small calculator is set at 10 microseconds, or one pulse every hundred thousandth of a second.

The ticking of the clock inside a calculator computer regulates a repetitive cycle of operations—the instruction cycle—that the machine performs continuously, over and over, as long as it is turned on. The instruction cycle is a two-stage affair.

On the first clock pulse the computer's clock unit—the switchboard—sends a message to memory, asking for the next instruction. The instruction (in the form of a binary pattern of charges) flashes back to the switchboard, which holds it until the next click of the clock. When the next clock pulse comes, the switchboard sends the instruction on to the appropriate part of the computer for execution. When one instruction has been executed, the controller waits for the next pulse. The clock ticks, the controller fetches the next instruction; the clock ticks again, and that instruction is carried out.

This two-stage instruction cycle—fetch and execute, fetch and execute—is the vital rhythm of the computer's life, as fundamental and as constant as the two-stage respiratory cycle—inhale and exhale, inhale and exhale—of the human body. The clock generator regulates the process so that each signal—from memory, from the control unit, from the keyboard, from anywhere—arrives at its destination before the next signal starts its journey through the circuit.

In DIM-I, as in video games and other simple digital tools, the sequence of instructions—the program—is permanently installed in memory when the machine is built. Such preprogrammed devices can do the exact tasks they were built for, and nothing else. A computer, in contrast, is not restricted to built-in programs; its versatility comes from the fact that it can be programmed by each user to perform a broad range of tasks. This is the difference between a "universal" machine—a computer—and a "dedicated" tool like a calculator. (There are pocket calculators that are called programmable, but in classic terms, machines like that are really computers, not calculators.) Buying a calculator is like buying a

ticket on the railroad; you can go only where the company's tracks
will take you. If you have a computer, on the other hand, you have
to steer it yourself, but you can drive it wherever you want to go.

In any case, the program, whether built in at the factory or provided separately, is a long chain of instructions that reside in a designated block of memory. In order to do any job, no matter how
mundane, the machine's central control unit has to fetch individual instructions from memory and execute them, one at a time,
one after another.

The fundamental fetch-and-execute cycle begins the instant
DIM-I is turned on. The clock generator starts generating regular
pulses, beeping out every 1/100,000 second. If you turn on a calculator and then put it down to sip some coffee or answer the
phone, the calculator may look as if it is doing nothing. In fact, it
is working furiously away, fetching and executing instructions—a
special instruction set that continually checks the keyboard, looking for input. All your digital devices—calculator, computer,
beeper, pager, cell phone, remote control, etc.—follow this same
special program when they are waiting for somebody to come
along and press a key. This special watching-and-waiting program
is called the idle routine. The idle routine is what most of our
modern tools are doing most of the time, during the periods when
they are not in active use.

DIM-I starts into its idle routine 1/100,000 second after it is
turned on, with the first clock pulse. *Tick.* The control unit goes
to the first location in the memory bank where instructions are
stored and fetches the first instruction: "Check the '1' key on the
keyboard." *Tick.* Now the controller executes that instruction.
That is, it sends an electronic query to the keyboard: "Has anybody pushed the '1' key lately?" The electronic reply comes back:
"No." With this sequence—the first instruction was fetched and
executed—the first instruction cycle has been completed.

Tick. The idle routine continues. The control unit goes back to memory and fetches the second instruction: "Check the '2' key on the keyboard." *Tick.* Having fetched and received an instruction, the controller now executes the order. It calls the keyboard: "Has anybody hit the '2' key lately?" The electronic reply: "No." *Tick.* The controller fetches the next instruction: "Check the '3' key." *Tick.* The controller queries the keyboard: "Anybody pushed the '3' key lately?" Reply: "No."

At this point, even the most primitive intelligence would start to appreciate that the drill here is to check all the keys seriatim to see if anybody has pushed. A child could figure out that the next two steps will be to check the "4" and the "5" keys. DIM-I, however, is much more stupid than any child. It has no intelligence; it is an ignorant digital mechanism. It knows nothing but the unvarying fetch-and-execute routine it has been programmed to follow mechanically.

Tick. Doggedly, the controller fetches the instruction in the next memory location. "Check the '4' key on the keyboard." *Tick.* Execute. *Tick.* Fetch. *Tick.* Execute. *Tick.* This is a computer's life. One by one, the control unit will check each of DIM-I's eighteen keys. If none has been pushed, there's only one thing for this mechanism to do. It runs through the whole routine again, and again, and again. Any human, indeed any chimpanzee, any creature with intelligence would be driven to distraction after the first few repetitive cycles of this. A digital device, in contrast, races its way through billions and billions of fetch-and-execute cycles every working day, never getting bored, never getting tired. It's a machine, after all, and that's what machines do for us. They take over the jobs that are too heavy, or too boring, for humans to manage.

At some point, every million or billion instruction cycles or so, DIM-I's unceasing vigilance pays off: Somebody pushes a key. The keyboard reports to the controller that a key has been pushed. But which key?

Each key on the keyboard of our calculator is a switch, just like the light switch on the wall. When somebody flicks the light switch on the wall, it lets current flow to the light bulb. When somebody pushes a key on the keyboard, it lets current flow to the control unit. This is just a quick surge of current—a swarm of electrons—and it is the same pulse no matter which key is hit. The "1," the "+," the "9"—every key sends the same pulse. (In those warehouse-size computers of the 1950s, some designers did try to vary the current for each signal, so that a "9" sent a surge nine times as powerful through the circuit as a "1." But this approach takes huge amounts of power and generates huge amounts of heat. Things are much simpler if you have each key send the same pulse. The only problem is that the control unit needs to discern whether a given pulse is supposed to be a "1," a "+," or a "9." That's where the clock comes in.)

The control unit at the heart of DIM-I is like the stationmaster at some isolated depot along the main freight line. The stationmaster knows that four trains come through from the south each day: the 10:00 A.M. from Tulsa, the noon train from Natchez, the 2:00 P.M. from Texarkana, and the 4:00 P.M. from Fort Worth. His job is to watch for the trains and switch each one to the right track to reach its destination.

Each morning at about 9:45 the stationmaster hears the remote rumble of a train chugging in from the south. He can't see the train, and all trains sound alike at a distance. Yet, the station man knows that this is the train from Tulsa. It has to be; it's the only train scheduled for that hour. He checks off the Tulsa train in his logbook; then he reaches in the desk drawer and fetches his stationmaster's manual to get the instruction for which track he should route it to. Then he gets a cup of coffee and goes back into his idle routine. At 11:45 he hears another engine; the stationmaster can't see that, either, and all trains sound alike. But he takes out the logbook and checks off the train from Natchez. He knows it has to be the train from Natchez, because the timetable tells him so.

So it is for the control unit inside DIM-I. All through the idle routine it has been querying each of the eighteen keys, one at a time. Each key is queried at a specific time interval—intervals determined by the steady heartbeat of the clock generator. The control unit can receive a pulse from a given key only when it is addressing that particular key. If a pulse comes in from the keyboard during the 2/100,000 second when the control unit is querying the "3" key, then the pulse had to be a 3. Somebody just pushed the "3" button. Pulses from the keyboard, like locomotives at a distance, all sound alike. But the control unit knows this pulse had to come from the "3" key, because the timetable tells it so.

Except that there's one other possibility: The surge of current may not have come from the keyboard at all. It could have been nothing more than random electronic "noise." Our atmosphere is almost always noisy with stray electromagnetic pulses, some generated by nature, others by mankind's machines and weapons. Weapons, particularly nuclear weapons, are prolific producers of electromagnetic pulses, or EMP, to use the Pentagon acronym. One of the many unknowns about the world's nuclear arsenals is the quantity of EMP that might result from an atomic blast or test. It is considered possible that any nuclear weapon, once triggered, would release so much electromagnetic "noise" that the control units of the world's computers—including the computers supposedly directing strategic planning in war rooms around the planet—would be flooded with stray pulses and would break down in a state of terminal confusion. Even in time of peace, though, there is enough EMP floating around that every digital device must be programmed to check the signals coming in. Before the control unit of DIM-I can really be sure it got a signal from the keyboard, therefore, it goes back to the "3" key a few more times to make sure the key has actually been pressed. This and execute, in precise rhythm with the ticking of the clock, the instructions to operate the "output decoder" circuitry—the series of logic gates that will transform the binary 0011 that has been stored in the register into the digit 3 on the display screen.

Small digital devices—watches, calculators, cell phones, car navigation systems, etc.—can use any of a variety of display techniques. The choice in any particular machine is dictated by a number of considerations—the amount of power available, the type of information that needs to be displayed, the speed at which the image will change, etc. For a cheap little handheld calculator like DIM-I, the only two realistic possibilities—because they work fine on battery power—are light-emitting diodes and liquid crystal displays. The light-emitting diode (known to engineers as LED) system employs a group of small diodes made of a special semiconductor material that work something like a standard light bulb. They give off a bright glow—red, yellow, green, orange, or white, depending on the material used—when current flows through. Unlike a light bulb, though, an LED can be made to glow with a minute amount of current; and since it has no filament to heat up, the LED gives off light without heat. (Although it doesn't have much to do with this chapter, I can't resist pointing out that, early in the twenty-first century, the glowing-filament light bulb that Edison invented in the nineteenth century will finally be replaced. Engineers have been developing brighter and whiter LEDs; soon a white LED will replace the familiar glass light bulb as the standard home lighting source. These new lights will use far less power, they won't get hot, and they'll never burn out. So don't buy any light bulb stock.) The other standard display device for low-powered devices is the liquid crystal display (LCD), a creamy gray liquid that works the opposite of a light bulb: When current is applied, the liquid crystal blocks the flow of current and thus turns black. The result is the familiar black digit floating on a gray background that you see whenever you look at your digital watch. Whether it uses LED or LCD, though, the display produces the characteristic squared-off digits that have become part and parcel of the microelectronic age, staring out at us everywhere from the digital readouts of clocks and calculators, scoreboards and scales.

Each of the ten digits from 0 to 9, together with a few other standard symbols such as the minus sign for negative numbers, can be created in the familiar squared-off style from a rectangular cluster containing seven separate diodes, arranged this way:

Drawing all the digits using just these pieces turns into a sort of high-tech version of the maddening little puzzles that ask you to form a "T" out of four weirdly shaped pieces of plastic. In a seven-segment display, some shapes are simpler to make than others. To make a number "8," all seven diode segments would be lighted. A "1" is formed by lighting only segments b and c. A "5" can be made with segments a, f, g, c, and d:

If an eighth diode, shaped like a period, is added to the cluster, this eighth segment can represent a decimal point:

The calculator's output circuit has to solve the puzzle: it has to draw the decimal digit representing whatever number the control unit sends along. The output unit consists of a group of logic gates that sends out a pattern of signals to selected segments of the display. In the case at hand, the binary pattern 0011 flows from the control unit to the output gates. The gates, in turn, will turn on segments a, b, g, c, d, and h, but not segments e and f. The display, consequently, lights up like this:

Moving the "3" from the keyboard to the display screen—and writing it down in a register along the way—has taken hundreds or thousands of instruction cycles. If the machine took just one second to fetch and execute each instruction, we would have waited ten or fifteen minutes for the display to light up. Anybody who has surfed the Internet using a standard telephone line knows what it feels like to sit around waiting for something to show up on the screen, but our little calculator doesn't make anybody wait. With its clock ticking every hundred thousandth of a second, DIM-I runs through all the instructions and completes the job in a total time of about .005 second. On the human scale, this is effectively instantaneous: the digit appears to hit the display screen the instant we push the key. The calculator is quicker than the eye, even though it turns every task into a ridiculously complicated series of fetch-and-execute operations.

It's also quicker than the hand. Presumably, the human operator will soon be punching some more keys. But even an unusually nimble human finger will take two or three tenths of a second between keys—virtually eons of time on the microelectronic scale. Thus the control unit, even at the unhurried (by microelectronic standards) pace of our 10-microsecond calculator, has more than enough time to check for EMP, write the "3" in the register, send it on to the output circuitry, and return to its old familiar pastime—the idle routine—before the finger can move on. Indeed, a calculator will check the keyboard thousands of times in the interval when the human finger is between two key punches.

On its next few thousand queries to the keyboard, consequently, control will probably find a finger still resting on the "3" key. This manifestation of human torpidity requires a whole new subgroup of idle routine instructions so that DIM-I can figure out what the user really has in mind: was she just taking her time about punching that 3, or is she really trying to enter a number like 33, or 333333? Among programmers, in fact, there's endless dispute about how long a control unit should wait before deciding that the finger on the "3" key is sending a new 3. The same problem shows up on cellular phone keyboards. Some people, particularly teenagers who send endless text messages on their phones, have become amazingly fast typists on that tiny keyboard. The phone has to be programmed to deal with both whiz-kid speedsters and slow-poke adults who punch out each new number slowly, slowly, slowly.

Once control figures out how many 3's were intended, it can go back to its stationmaster duties: watching the keys, checking each pulse it senses against the timetable, logging it in, and sending the appropriate signals to various parts of the circuitry. It may switch the pulse onto a wire leading to the main memory unit, to a temporary memory register, or to the display screen.

And occasionally it sends the signals on to the most important destination in any digital machine. This is the complex nest of logic gates that actually does the calculating. It is known by various names: arithmetic-logic unit, adder-subtracter, etc. Whatever it's called, this central arithmetic circuitry is the place where a computer does its computing.

Let's assume, for example, that the person who pushed DIM-I's "3" key was starting to punch in an addition problem: 3 + 2 = . As we have seen, the first keystroke resulted in a byzantine sequence of operations that left 0011 (binary 3) in a memory register and displayed a "3" on the screen. For the next two keystrokes, the same general sequences are followed again, in all their labyrinthine complexity, as DIM-I digests the "+" and the "2."

Things become even more complicated when the pulse comes in from the "=" key. That pulse is the call to action. As soon as control senses that someone has touched that key, it jumps to a new set of instructions that dispatches the number and the plus sign to the arithmetic unit. The most complicated part of any digital device is the circuitry for that unit. All the switches and all the wiring must be arranged so that no matter what binary pulses are fed in, only the correct answer can come out. It is only because humans have figured out how to build arithmethic-logic circuitry that an ignorant tool can perform logical and arithmetic operations.

The basic building blocks of the arithmetic circuit in any digital tool, big or small, are the standard logic gates that Claude Shannon first proposed in 1937. As Shannon showed then, various combinations of switches can be wired together with a few resistors and capacitors to carry out the basic logical operations that George Boole had formulated. These logic circuits are called gates; the gates take in and send out impulses according to established rules. The simplest of these logic gates is the circuit called a NOT gate—a set of switches with an output that is always the opposite of the input. That is, if current flows into the NOT gate, current does not flow out. When current is not flowing in, current does flow out.

Translating this to binary math, a NOT gate that takes in a "1" will send out a "0," and vice versa. The complete operation of this small, simple circuit can be set forth in a small, simple logic table:

In	Out
1	0
0	1

Because the operation of this circuit is a function of fundamental physical forces, there can be no other results than those shown in the table. Accordingly, a table like this is known to computer engineers as a truth table.

The standard approach to circuit design is to prepare the truth table that provides the desired result and then, using Shannon's techniques, build a circuit that produces a result to match the table. It is simple enough, for example, to produce the truth table for the Boolean AND operation. As we saw in Chapter 6, the rule of the AND operation—typified by the sleepy commuter who has to get out of bed in the morning if the clock says it is 8:00 A.M. AND the calendar says it is a workday—is that the result is yes only if condition A AND condition B are both true. To state that logical rule in the binary terms that DIM-I can understand, the output is 1 only if input A AND input B are both 1. If either input is not 1, then the AND gate has to emit a 0. Set it down in a truth table, and the AND operation looks like this:

Input A	Input B	Output
0	0	0
0	1	0
1	0	0
1	1	1

That table is a little more complicated than the simple NOT table we just saw. And it is possible for these things to become a whole lot more complicated. An AND operation, for example, could have three or four or eight different inputs. The basic logical AND rule is the same—the output is 1 only if input A AND input B AND input C AND input D, etc., are all 1—but the truth table gets more complicated to draw and to replicate in electric circuitry.

Further, different logic gates can be chained one after another in different combinations to achieve different logical results. In a common instance, a NOT gate is wired to the output of an AND gate, so that the result of the AND operation is directly reversed. This combination of NOT plus AND creates a new logic gate called a NAND gate. For various reasons, this has become the most popular building block of all for many computer designers.

To make a mindless machine like DIM-I add two digits and come up with the right answer every time, then, the circuit designer first draws up the truth table for a two-input addition problem. Since digital machines do their arithmetic in binary numbers, the truth table encompasses all the rules for binary addition:

1st digit	2nd digit	Sum
0	0	0
0	1	1
1	0	1
1	1	10

(The 10, which is read "one-zero," is the binary version of 2.)

Since people started building digital computational devices, standard circuits have been developed to replicate every necessary logic gate and every necessary mathematical operation. Engineers have perfected several different circuits that will produce precisely this binary result. The necessary patterns of switches are built into the corner of a microprocessor chip that does logic and arithmetic. Since DIM-I is such a simple device, it uses the simplest addition circuit there is, constructed with about a dozen standard gates. The electronic pulses flow into the gates in patterns representing the binary digits to be added. What comes out is a pattern of pulses representing the sum of those two digits—and a carry digit, if necessary, to be carried over to the next column.

And this circuitry comes into use when the control unit inside DIM-I notices that somebody has pushed the "=" key. Control now goes back to its temporary storage registers to find what the previous key clicks were so that it can do the computation. The registers record that somebody pushed the keys "3," "+," and "2." So DIM-I now faces the formidable task of adding 3 + 2 and finding the sum. To do so, control feeds the 3 and the 2 (in binary form, of course) into an addition circuit, one column at a time:

$$\begin{array}{r} 0\ 0\ 1\ 1 \\ +0\ 0\ 1\ 0 \\ \hline \end{array}$$

Starting with the right-hand column (the 1's column of the binary number), each pair of digits enters the logic gates. Because of the way the gates are wired, there is only one pattern of pulses that can possibly emerge:

$$0\ 1\ 0\ 1$$

(That's a binary 5, of course.)

Now that the addition circuitry has emitted an answer, the con-

trol unit gets back into the act. It sends a signal to the output circuitry: "Get ready to display the answer." The binary pulses flow
in; current flows out along selected wires leading to the display.
The chosen segments light up, and the binary answer to the problem, 0101, is translated on the screen to its decimal equivalent:

Ta da! In the most roundabout conceivable fashion, DIM-I has
actually solved a problem. In the minute step-by-step manner of
the microprocessor, the seemingly simple task of adding 3 + 2 has
become a Cecil B. DeMille production, with thousands of extras
following the director's instructions to the letter. A problem any
human mind could work in a flash had to be broken down for the
calculator into thousands and thousands of individual instructions that led the machine by the nose through its paces.

Watching the way a computer has to go back to the ridiculously
detailed instruction manual to check on everything it does reminds me of that wonderful scene with the two evil scientists in
the Beatles' movie *Help!* One white-coated scientist makes a mis-
take, infuriating his boss. "From now on," the boss roars, "don't do
anything without checking first with me." "Yes, boss," the under-
ling replies. Now the boss says, "Okay, come over here!" The un-
derling starts to walk across the room. Then he remembers the
rule that he has to check everything first with the boss. "Can I
move my left foot?" he asks. "Can I move my right foot?" And so
on. Naturally, this makes the boss even more impatient.

But of course, DIM-I is too ignorant to be patient or impatient.
It is a tool, like a hammer or a screwdriver, and the only way it can
do anything is to follow the precise instructions humans have
given it, to the most minute degree. That's why adding two num-

bers takes all those fetches and all those executes. It is a process that only a mindless mechanism could love. It is a process that only a powerful intelligence could have invented. But then, digital mechanisms were invented by the most powerful computer on earth — the human mind.

10

SUNSET, SUNRISE

A spectacularly rococo official document showed up in Jack Kilby's mailbox in Dallas on the morning of June 27, 1974. Framed in a silken strip of bright red ribbon, affixed with a gleaming gold foil seal, engraved in a florid cursive script, and illustrated with an ornate etching that depicted, among much else, the American eagle, the stars and stripes, assorted machinery, and the headquarters building of the U.S. Department of Commerce, the document poured forth its message in a rich flood of legalistic prose:

> TO ALL TO WHOM THESE PRESENTS SHALL COME: WHEREAS Jack S. Kilby; Jerry D. Merryman; James H. Van Tassel, all of Dallas, Texas . . . presented to the COMMIS-SIONER OF PATENTS a petition praying for the grant of LETTERS PATENT for an alleged new and useful invention the title and a description of which are contained in the specification of which a copy is hereunto annexed and made part hereof . . . and WHEREAS upon due examination made the said Claimants are adjudged to be justly entitled to a patent under the law: Now therefore these LETTERS PATENT are to grant . . . for the term of SEVENTEEN years from the date of this grant.

The specification thereunto annexed and made part thereof was, in fact, the patent application that Texas Instruments had sent off years earlier when Kilby and his "Cal Tech" team had successfully developed the first pocket calculator. With the formal receipt of the patent, TI could now replace the words "Patent pending" on its pocket calculators with the official patent number: U.S. Pat. 3,819,921 — a legend that is still being molded today into the plastic cases of calculators turned out by Texas Instruments and other firms it has licensed to copy the basic Kilby architecture.

In sharp contrast to the long legal struggle over the integrated circuit, there had been no contest about the patent for the handheld calculator. Everyone recognized that Kilby, Merryman, and Van Tassel were entitled to the honor. By the time the patent was issued, however, the honor of the thing was just about all it stood for. By the middle of 1974, it was clear that Texas Instruments, and for that matter all other American manufacturers, had lost control of the calculator business. This revolutionary product — conceived, born, and bred in Texas, U.S.A. — was becoming a Japanese preserve.

Given the Japanese dominance of other facets of the consumer electronics business, it was not altogether surprising that firms like Casio and Panasonic had caught and passed their American competitors. But it was still a bitter irony. By succeeding with the pocket calculator, the Japanese electronics firms had added a sour twist to one of the happier chapters in contemporary American business history.

Since its first appearance in the 1930s, the office calculator had been recognized worldwide as an American product. Despite impressive competition from European firms, U.S. manufacturers dominated the market. Then, in the 1960s, when a "calculator" was still a hefty, expensive, desktop apparatus, Japanese manufacturers had moved into the U.S. market and simply overwhelmed the traditional American office-machine firms. Japan's market dominance increased steadily, and the situation was bleak for the American side until the day in 1971 when Jack Kilby's first hand-held calculator went on the market. The application of microelectronics to the calculator business completely turned the tables. The chip-based calculator—smaller, cheaper, and more efficient than any Japanese product—put American labels back on top.

This reversal was so dramatic that the Commerce Department issued a detailed report on the American triumph: "With innovative products and aggressive pricing policies backed by high volume production efficiencies, U.S. firms rapidly regained control of the U.S. calculator market from their chief competitors, the Japanese." The report went on to say that this turnabout was "an excellent illustration of the way a free and competitive industrial system can provide benefits to the consumer," and a model that other U.S. industries could emulate in their struggles against foreign competition.

Unfortunately, the United States was not the only place with a free and competitive industrial system, and American firms were not the only ones in the calculator business that knew how to practice aggressive pricing and production efficiencies. By the time Jack Kilby received his letters patent, Japanese manufacturers had obtained the information they needed to make handheld calculators of their own. In part, they developed their own calculator know-how; mainly, though, they borrowed American techniques, either buying or appropriating patented processes from U.S. firms. (TI sued Casio in 1982 for infringement of the Kilby calculator patent; the case was settled with Casio paying a considerable sum.)

In any case, the competition from Japan was so intense and so successful that, two years after its original calculator report, the Commerce Department had no choice but to issue an unhappy update: "Since 1974, the situation has once again been reversed. . . . The departure of U.S. firms from the calculator industry continues. . . . Japanese-origin calculator imports have again been a major factor in this downward trend." When that was written, in 1977, the Japanese made about 45 percent of the calculators sold in the United States. By the early 1980s, Japanese firms' share of the U.S. market, in dollar value, was above 70 percent. (By the turn of the century, both Japanese and American manufacturers had moved on to higher value-added products, and most pocket calculators came from China or Southeast Asia.)

As the Commerce Department noted, the pattern of conquest in calculators in the seventies was familiar to anyone who had watched Japanese firms make equally successful forays into U.S. markets ranging from automobiles to zippers. The Japanese had first moved into the low-priced segment of the business, turning out simple, "plain vanilla" calculators that offered the consumer greater variety, equal or better reliability, and lower prices than comparable U.S.-made brands. Having established a beachhead at the low end of the market, the invaders moved to broaden their territory. "All they did was take our product and add a bunch of bells and whistles," a U.S. manufacturer complained. But the bells and whistles—not to mention the calorie counters, currency converters, perpetual calendars, astrology tables, biorhythm charts, and assorted other add-ons dreamed up by the marketing men in Tokyo—sold calculators by the millions. In the third stage of their assault, the Japanese set their sights on more sophisticated, higher-priced models that provided the highest profit margins. "This pattern of market activity," the Commerce Department's second report said, "has been observed for other products, such as television sets."

The comparison to television sets had to be a frightening one for anybody in the U.S. electronics industry. Television was the paradigm case of international competition in modern electronics, but it is hardly a pleasant paradigm for American industrialists to ponder. Just about every significant advance in the development of television has come from the United States. Yet the Japanese repeatedly trumped the American genius for innovation with their own skills at production and marketing.

The ancestry of the television set can be traced to the work of several late-nineteenth-century European physicists who developed and perfected the cathode ray tube. The modern TV is a direct descendant of this device, a lineage reflected in the slang terms "tube" and "boob tube." The cathode ray tube, originally built strictly as a laboratory apparatus, was a long, closed glass tube with a mechanism inside that shot a "cathode ray"—the mysterious beam of energy that J. J. Thomson determined to be a stream of electrons—down the length of the tube. At the opposite end of the tube the glass was coated with phosphor. An electron shot from the cathode gun raced down the tube and struck the phosphorescent glass at the far end. Precisely where the electron hit, the phosphor would glow with a bright point of light.

Peering at an interesting pattern of glowing dots on a cathode ray tube in 1908, the British physicist A. A. Campbell-Swinton hit upon the fundamental idea of television: the dots of light on the glass tube could form a pointillistic picture—the television image. (If you put down this book for a moment and look closely, very closely, at a television screen, you will see that the picture on the glass is actually composed of thousands of individual points of light, à la the French pointillist painter Georges Seurat.) Campbell-Swinton named his idea "Distant Electrical Vision"; the term "television" was apparently coined by the French. As every schoolchild in the Highlands knows, the Scotsman John Logie Baird pursued the idea with some success. But the real invention of television came about in the 1920s in a pair of American laboratories run by a pair of competing inventors who battled furiously to be first to perfect the concept.

In one corner was Vladimir Zworykin, an officer in the tsar's army who had fled Russia during the 1917 revolution and eventually landed in David Sarnoff's laboratory at RCA in New York. In the other was Philo T. Farnsworth, a fiercely independent farm boy from Idaho who would not sign on with any of the major electronics firms and was forced to scrounge constantly for money to support his research. (After years of work, and just on the verge of final success, Farnsworth retired back to the farm at the age of thirty-four—a classic burnt-out case.)

Spurred on by the whip of competition with each other, Zworykin and Farnsworth rapidly surpassed the efforts of Baird and other Europeans, and by 1929 they had developed, between them, a camera that could convert a visual image into electronic signals and a picture tube that could translate these signals back into the visual image. The heart of the Zworykin/Farnsworth TV—and of almost all television sets today—is a cathode that fires a moving beam of electrons toward the phosphorescent screen. The electron gun sprays a picture on the screen the way an aerosol paint can sprays pictures on the subway wall. The graffiti, of course, is on the wall for good; the television tube, in contrast, sprays a new image on the screen sixty times each second, creating the illusion of a moving picture.

Serious commercial development of television was delayed until after World War II, when the U.S. government finally got around to issuing technical regulations for frequencies, bandwidths, and the like, so that all stations could broadcast the same standard signal. In the first year after the war, 6,000 television sets were sold in the United States. In 1948, Americans bought 1 million sets; two years later, 7 million. The boom was on. Because American technology set the pace, the U.S. standards were adopted, with minor variations, around the world.

As their sales skyrocketed, U.S. television manufacturers developed a marketing structure not unlike that of another booming American business, the automobile industry. The major television makers—Admiral, DuMont, Sylvania, etc.—sold their products through franchise dealerships that generally handled only one brand. The dealers provided factory-authorized repair, an important function at a time when the family TV was likely to be in the shop as often as the family car. And they dutifully pushed the large console models that brought manufacturers the largest profit margins.

This retail structure made it doubly hard for new firms, particularly overseas manufacturers, to move into the U.S. market. It was the biggest and most lucrative television market by far, but there was no way for an outsider to break into the dealership system, which was controlled by the big domestic firms. Further, without a network of parts depots and factory-trained repairmen, how could an overseas competitor keep his sets in working order?

These problems persuaded the major European electronics firms to stay home. But the Japanese manufacturers, prodded by government officials eager to increase exports, took a long look at the United States in the early 1960s and decided to give it a try. On technical matters, the Japanese initially copied what others had done. The firms purchased the rights to more than 400 patents from U.S. and European firms and generally waited for others to make the important breakthroughs. In production and marketing, however, Japan did the innovating.

Japanese television firms approached the U.S. market with a wider variety of models than many domestic makers offered, including a line of smaller table-model sets and the first portables. By targeting the low-price segment of the market, they were able to sell their sets through the fast-growing discount chains, circumventing the traditional dealer network. To deal with the lack of trained repairmen, the Japanese eliminated the need for repair. They honed a highly disciplined labor force in their factories and introduced extensive quality-control mechanisms, so that Japanese television sets would work reliably without constant maintenance.

The results were dramatic. In 1961 the six Japanese manufacturers combined sold less than one half of one percent of the black-and-white television sets purchased in the United States. They moved up to 2 percent of the market in 1962 and 13 percent in 1965; thereafter, the percentage kept going up, year after year.

American manufacturers shrugged off this challenge. By the mid-sixties, they were concentrating on an even hotter new product—color television, in which U.S. technology led the world by a large margin. But the color TV market turned into a rerun of the monochrome story. The Japanese aggressively purchased patent licenses in the West. They entered the fray a few years behind the American firms but made up for the late start with their familiar combination of extensive variety, low prices, and high quality. By 1976, Japan was the world's leading producer of color television sets and had 35 percent of the U.S. market.

At this juncture, some U.S. firms simply gave up. Admiral quit the business. Motorola sold its television operation to Matsushita. Others fought back, both in the market and in Washington. U.S. manufacturers charged that some Japanese competitors were "dumping" TV sets in the United States at predatory prices; under the law, this could have resulted in multimillion-dollar fines for the Japanese firms, but diplomatic considerations prevented that result. Instead, the governments of Japan and the United States in 1977 negotiated an "orderly marketing agreement"—the diplomatic term for an import quota. Those yearly agreements preserved almost 80 percent of the U.S. market for domestic makers (the term "domestic," in this context, included Japanese firms building sets in the United States). In the end, it was diplomacy, not technology or business acumen, that saved the United States a piece of the action in a market it had pioneered.

The television saga would be an unsettling one from the American point of view even if it were the sole instance of such an abrupt industrial defeat. But in consumer electronics, as in automobiles and steel, the history just related was the rule, not the exception. Throughout the final decades of the twentieth century, American firms were overtaken by Japanese manufacturers offering a broadly appealing range of products that combined premium quality with competitive prices.

The Commerce Department, displaying the universal bureaucratic yen for numbers and tables, issued a study in 1980 that reduced the whole sorry state of affairs to a single chart:

IMPORT PENETRATION — CONSUMER ELECTRONICS

Product	Imports as % of U.S. sales
Videotape equipment	100
Household radios	100
CB radios	90
TV—monochrome	85
Electronic watches	68
Hi-fi, stereo components	64
Tape recorders (audio)	35
Microwave ovens	25
TV—color	18

In each case, the product was conceived and developed in the United States; in each case, the Japanese (sometimes followed by producers from other nations, such as Taiwan, South Korea, and, more recently, China) borrowed the technology and then swept into the American market.

The American semiconductor industry, during the first booming decade after Jack Kilby and Bob Noyce hit upon the monolithic idea, could safely look upon these developments in consumer electronics as irrelevant. In chips, the American pioneers had an enormous technological lead over foreign competitors everywhere, European or Asian. And however daunting the Japanese might be in consumer electronic goods, the chip was a different animal. The Japanese had proven adept at moving into relatively stable technological fields, where the design of the product was settled, and producing large volumes of goods at reasonable prices. But the integrated circuit business, in its first quarter century, was never stable. Improvements came so rapidly that there was no such thing, really, as a settled design; prices fell so rapidly along the learning curve that undercutting from foreign competitors did not loom as a serious threat.

Still, there were signs during the 1960s and early 1970s that the Japanese had a collective eye on the semiconductor business. The big Japanese electronics firms—outfits like Fujitsu, Nippon Electric Company (NEC), Hitachi, and Sony—began buying licenses for every U.S. semiconductor patent they could find; between 1964 and 1970, royalty payments from Japan on U.S. semiconductor patents rose by a factor of ten, from $2.6 million to $25 million per year. For some U.S. firms, particularly Fairchild, this became an important source of no-risk income. "American firms have generally been very cooperative," a Brookings Institution economist, John Tilton, reported. "With few exceptions, they have been willing to license Japanese as well as other foreign firms and aid them in assimilating new semiconductor technologies, even though in the process they are helping establish potential rivals."

There was only one major U.S. firm that refused to cooperate:

Texas Instruments. TI spurned royalty payments and set a more ambitious price on its semiconductor patents: no Japanese firm could use them unless the Japanese government permitted TI to set up manufacturing operations in Japan. This demand posed a serious dilemma in Tokyo. The Japanese were determined to exclude most foreign competition in high-tech fields from their country. But in order to produce integrated circuits, the Japanese firms would need access to both the Noyce patent from Fairchild and the Kilby patent from Texas Instruments. Fairchild was willing to sell its technology, for a royalty of 4.5 cents on every dollar the Japanese makers earned on chips. But the Texans hung tough. In 1968, after several years of offers and counteroffers, the Dallas firm was finally permitted—in return for a license to the Kilby patent—to open a Japanese plant. For more than a decade thereafter, TI was the only U.S. semiconductor producer with any significant sales in Japan. Not until the late 1980s, when the Reagan administration finally slapped an import ban on Japanese memory chips, did Intel and other U.S. firms manage to open plants and design facilities in Japan.

Whether or not the American firms were aware of it, both the vigorous pursuit of U.S. patent rights and the exclusion of U.S. competition were part of a grand strategy devised by the Japanese government to win that nation global preeminence in chips. The semiconductor business was one of several high-tech industries targeted by the Japanese Ministry of International Trade and Industry (MITI) for intensive development. In its ten-year plan, or "vision," for the 1980s, MITI concluded that microelectronics was perfect for the island nation, because it required large quantities of human resources, such as advanced engineering and diligent workers, which Japan has in abundance, but only small amounts of energy and natural resources, which Japan lacks.

Officially, MITI's decisions about what was good for Japanese industry constituted nothing more than "administrative guidance," which companies were legally free to ignore. Some did. In the 1950s, for example, someone at MITI had the brilliant insight that Europeans and Americans would never shell out their money for cars bearing names like Honda or Toyota. MITI issued "guidance" telling a group of automakers to get together and design a single "people's car" that would represent the entire Japanese industry around the world. Honda, Toyota, Nissan, etc., rejected this idea, with stunning worldwide results. When MITI worked with its electronics industry in the sixties to plan Japan's foray into the low-priced end of the U.S. television market, one manufacturer, Sony, spurned the consensus view and found a lucrative niche of its own, as a prestige upper-bracket label.

As a rule, though, MITI's policies tend to become industrial practice. They certainly did in semiconductors. Just as MITI planned it, Japanese electronics firms acquired the U.S. technology required to get started in the manufacture of integrated circuits in the mid-sixties. Just as MITI planned it, these firms were able to rely on sales to domestic Japanese computer and telecommunications firms—free of competition from any U.S. firm except Texas Instruments—to provide the financial cushion necessary for an assault on the world market for chips. And just as MITI proposed, the five largest electronics firms banded together in the early 1970s for a cooperative research endeavor—funded partly by the government and partly by the companies—to develop manufacturing techniques for very-large-scale integrated (VLSI) chips.

For all this effort, however, the Japanese electronics industry still lagged far behind American firms in most areas of microelectronics. The one product in which the Japanese took any significant market share was random-access memory, or RAM, chips. The memory circuit, pioneered by Robert Noyce and Gordon Moore at Intel in 1968, was nicely suited to the Japanese strengths: a simple and relatively stable product that was consumed in huge quantities by computer manufacturers around the world. Even in the RAM market, though, U.S. dominance was unchallenged until the middle of the 1970s. Then the American industry gave Japan its chance.

During the prolonged recession following the 1973 oil embargo, American semiconductor firms reacted the way American manufacturers normally react to recessions: they laid off workers, closed plants, and generally hunkered down to await an upturn. In 1976, when the economy came roaring back, there was an enormous burst of demand from computer firms for what was then the most advanced memory chip: the 16K RAM, capable of storing some 16,000 bits of information. The U.S. firms could not rebuild fast enough to meet the need. Their customers went shopping for an alternate source of RAM chips—and found it in Japan.

Following standard Japanese industrial practice, the big Japanese electronics firms had maintained their work force and their production capacity during the recession. They absorbed the costs of full employment during slack times on the theory that the investment in keeping a trained, loyal work force on the job would pay off when things turned up. They were, accordingly, in the catbird seat when the market for 16K RAMs took off in the mid-1970s. Silicon Valley's inability to meet demand gave Japan a golden opportunity to show the world what it could do, and Japanese firms leaped at the chance. They began dispatching high-quality, competitively priced chips around the world. By 1980 the Japanese had 42 percent of the world market in memory chips, a market that American firms had once owned.

More important, Japanese firms had put themselves in perfect position to compete when the next generation of memory chips, the 64K RAM, was developed in the late 1970s. This time Japanese firms had no technological lag to worry about, and this time they had established markets everywhere for their memory chips.

After two years of nip-and-tuck competition, Japanese firms finally eclipsed the Americans. By 1983 they controlled well over half the world market for 64K RAMs.

The Japanese success in this one product line prompted all manner of distressed breast-beating in the United States—far more than the situation actually warranted. The RAM memory chips constitute a high-volume part of the semiconductor market but not a particularly important or remunerative one. In dollar terms RAM chips probably represent somewhere between 5 and 10 percent of annual semiconductor sales. Americans have maintained supremacy in virtually every other type of integrated circuit. For Silicon Valley to worry about Japanese or Korean or Taiwanese sales of memory chips is like General Motors losing sleep over a smaller company that has done well in the spark plug business.

But many Americans were worried. The Japanese inroads in chips, a congressional committee reported, "indicate the potential for an irreversible loss of world leadership by U.S. firms in the innovation and diffusion of semiconductor technology." The Silicon Valley firms were sufficiently alarmed to form a trade group, the Semiconductor Industry Association, specifically to fight off the foreign challenge. "The television people woke up when the Japanese had 20 percent of the market, and went to the government when the Japanese had 40 percent," the group's executive director told *The New York Times*. "That's a little late."

The Semiconductor Industry Association began turning out a steady flow of studies and brochures and advertisements, along with petitions to various government commissions and agencies, asserting that the competition from Japan was unfair. The SIA cited MITI's determination to keep U.S. companies out of the Japanese market. It complained about Japanese government contributions—something over $100 million—to the very-large-scale integration research and development venture. It pointed out that

the Japanese firms, partly because of that nation's industrial structure and partly because of support from MITI, received loans
from Japanese banks at much lower interest rates than any electronics firms could hope for in the United States.

"I just don't want to pretend I'm in a fair fight. I'm not," wrote
Jerry Sanders, chairman of Advanced Micro Devices, in a statement that crystallized the SIA position. "Do you know how the
Japanese got the dynamic RAM business? They bought it. (If I
had their deal, I'd have bought it too.) They pay 6 percent, maybe
7 percent, for capital. I pay 18 percent on a good day. . . . They
start every product development cycle with hundreds of millions
of dollars of free R&D every year, paid for by their government.
Good for them. But then their parts arrive here in a flood."

There was, however, another point of view—a view held,
among other places, at Texas Instruments, which never found reason to become a member of the Semiconductor Industry Association. The dissenters pointed out that despite Japan's efforts to
exclude foreign competitors, American firms had always had a
larger share of the Japanese semiconductor market than the Japanese had gained in the United States. They argued that the American semiconductor industry, launched on a wave of government
financing and still receiving tens of millions of dollars annually
from the Pentagon for research and development, was hardly in a
position to carp at MITI's grants to Japanese firms. The SIA's
American critics also recalled that the Silicon Valley firms, by
selling their patents and by cutting capacity just before the 1976
boom, were themselves partly responsible for the Japanese success. "Those fellows on the West Coast sort of have schizophre-
nia," Fred Bucy, the president of Texas Instruments, said after the
SIA was founded. "They had the same leverage as we did. . . . But
they were very shortsighted in the way they handled the patent situation."

There was one other factor, too, fueling the Japanese success in

RAM memory chips, but it was something that American trade groups were not eager to talk about. It was a familiar factor to Americans who had observed the Japanese success in television, in cameras, in automobiles—the quality factor.

In the middle 1970s, when the Japanese electronics firms first managed to break into the American semiconductor market, U.S. companies buying imported memory circuits for installation in their computers began to notice something interesting: Japanese chips were better. There was no discernible difference in performance, because all memory chips are built to work the same way. But there was a marked difference in reliability. Chips made in Japan were less likely to fail than the American product.

At first, the Japanese edge in quality was the dirty little secret of Silicon Valley. Hardly anybody talked about it, and when the subject did come up, American manufacturers heatedly denied that Japanese firms were turning out more reliable chips. That changed one morning in March 1980 when an American computer executive named Richard W. Anderson stood up at an industry meeting in Washington, D.C., and delivered a paper that came to be known as "The Anderson Bombshell."

Anderson was a division manager at Hewlett-Packard, a giant California-based manufacturer of electronics instruments and computers that is one of the world's biggest consumers of integrated circuits. It was he who had decided, rather reluctantly, to start buying Japanese memory chips for Hewlett-Packard computers, and at the Washington meeting he told his story.

We first introduced semiconductor memory in our computers in 1974. We got all our memory from United States suppliers. Then in 1977 the 16K, or 16 thousand-bit, RAM began to make its appearance. The first introductions that I was familiar with came from U.S. suppliers, and we hurried to implement this design into our product line. . . .

However, some months after introduction, the U.S. suppliers that we had been working with found themselves unable to meet our quantity demands, either due to yield or capacity problems, and this left us between the proverbial rock and a hard spot. So, after much anguish, we decided to talk to a Japanese company who had been calling on us telling us of their memory for some time. And I would like to state at the outset we took a very cautious approach because we remembered well the impressions from post–World War II Japanese products; namely, that they were cheap, low cost, and low quality. And so our engineers went through a very rigorous qualification program; and we were pleasantly surprised to find they qualified.

Anderson went on to say that, over time, he bought more and more chips from the Japanese firm. Although the fact was not immediately obvious, he said, Hewlett-Packard gradually began to realize that there was a significant difference in the Japanese memory circuits. "We had fewer failures in incoming inspection; we had fewer failures during production cycle; we saw fewer failures of products in customers' hands. . . . Not only was the quality good, but [it] was actually superior to what had been our experience with the domestic suppliers.

"Then came 1979," Anderson went on, "and a real market crunch hit the memory suppliers, particularly the U.S. manufacturers . . . and we found ourselves in short supply. So we went back to Japan and qualified two more Japanese suppliers for the product line that I'm responsible for. And again the same experience: excellent quality." Eventually, Anderson added, Hewlett-Packard compiled performance records on some 300,000 memory chips, of which half came from the Japanese suppliers and half from American makers. The final standings showed that all three Japanese firms were delivering higher quality goods than the best American manufacturer. "So that's a remarkable, and I would think to American suppliers, perhaps a frightening set of statistics," Anderson said.

Frightening it certainly was. The message that foreign competitors were outperforming the United States in semiconductors, the symbol of American technical preeminence, was a slap in the face that could not be ignored. The Anderson Bombshell, widely reported and corroborated by some other computer firms, made it impossible for American firms to deny any longer that there was a quality difference. Instead, they set out to learn how the Japanese had attained it. "U.S. Microelectronics Firms Study Japan for Secrets of Quality and Productivity," read a headline in *The Wall Street Journal* in 1981. The spate of books that appeared in the early 1980s extolling Japanese management practices became required reading in the semiconductor business. The American Electronics Association did a booming trade in seminars on Japanese quality control. Companies dispatched fact-finding teams to Tokyo to uncover the Japanese quality secret.

There was a strong streak of irony in this reaction. As the Japanese knew perfectly well, their quality mechanisms were hardly secret, particularly to Americans. After all, Japan had learned virtually all it knew about quality control from an American. The teacher was a federal bureaucrat, an obscure fellow, unknown in his homeland but recognized and respected throughout the Japanese islands. He was, like Jack Kilby and Robert Noyce, a son of the prairie.

W. Edwards Deming was born in Sioux City, Iowa, in 1900. Growing up on the prairie, young Deming, like many intellectually gifted youngsters, displayed talent both in mathematics and music. Eventually, the lure of math and science prevailed, and as an undergraduate at the University of Wyoming, Deming concentrated on math, physics, and engineering. He went on to graduate work at the Colorado School of Mines and at Yale, where he took a Ph.D. in physics in 1928. Dr. Deming then began a long career in federal service, first as a scientist in the U.S. Department of Agriculture's Bureau of Soils and later, when statistics became his dominant professional interest, at the Census Bureau. Eventually he left the federal service and began an extraordinarily vigorous lecturing and consulting business that he ran from the cluttered basement of his Washington home.

When I first met Dr. Deming he was past his eightieth birthday and traveling on business about 300 days per year, to Europe, Asia, and various corners of the United States. He was a tall, crusty, imposing gentleman with a fluffy white crew cut, spectacles, and a jutting nose that gave him the look of a tortoise poking around outside the shell. He told me that he had developed a method of squeezing two workdays out of one: he would work until midafternoon, nap until eight-thirty at night, and then rise for several more hours of business. He found time, naturally, for music—sessions at the piano with his wife, Lola Shupe Deming, and their children and grandchildren. He composed liturgical music; the Deming oeuvre includes two masses and a number of canticles. Otherwise, he was all business. I asked him once—he was eighty-three at the time—if he ever took a vacation. "Well, I'm not doing anything for the next few hours," came the gruff reply.

A nine-page "list of principal papers" that Deming provided his consulting clients reflects the changing focus of his interests over the years. The first dozen or so of his publications—many signed jointly by W. Edwards Deming and Lola Shupe—deal with physics and chemistry: "Equipotential surface electrons as an explanation of the packing effect"; "Note on the heat capacity of gases at low pressure." But midway through the 1930s the titles begin to reveal a developing interest in probability and statistics: "On the frequency interpretation of inverse probability"; "On a least squares adjustment of a sampled frequency table when the expected marginal totals are known"; "On the elimination of unknown ages in the 1940 population census."

This shift in Deming's scholarly output resulted from his growing appreciation for the enormous power of statistics to explain precisely what was going on in any repetitive process, including mass production of commercial goods. Deming learned that the statistician could act as a skilled detective when things went wrong in an industrial operation, searching his data to pinpoint the problem and eliminate it. He put this idea on paper in a seminal monograph, written in 1934 and updated thereafter, titled "On the Statistical Theory of Errors." The "Theory of Errors" was Deming's explication of the ideas of Dr. Walter Shewhart, the American mathematician who first applied statistical methods to the control of factory operations. Deming set out to be the evangelist of the Shewhart gospel, which held that careful statistical records were the essential key to consistent quality in manufacturing.

In the lectures he delivered around the world, and in his basic text, *Quality, Productivity, and Competitive Position*, Deming kept on preaching the lessons he had set forth back in 1934 in the "Theory of Errors." He stressed that quality and productivity result from diligent observance of certain fundamental rules—he has gathered them into a table of "14 points"—all of which can be summarized in a single, golden rule that should govern all work: "Do it right the first time."

Reduced to those six words, the Deming message seems obvious—a point the professor never failed to drive home. "It's so simple," he said in his lectures. "It's so obvious." But to carry out the obvious required detailed effort. To achieve consistent quality, those involved in any operation, from manufacturing semiconductors to managing a baseball team, must maintain "statistical control"—that is, careful, regular measurement of all aspects of the job. "By describing statistically exactly what is done," Deming said, "the method locates your problems and leads to innovations that solve them." When a manufacturing problem occurs, the quality control officer should probe the statistics like Sherlock Holmes probing the body at a murder scene.

One time, for example, Deming was retained by a shoe company that had run into costly but inexplicable manufacturing delays. Management was baffled; the same work force in the same factory had suddenly fallen far behind its normal production rate. Poring over time cards, maintenance charts, purchasing records, and the like, Deming found that someone had recently ordered a new brand of thread. Now it was all so obvious. The new, cheaper thread kept breaking, forcing workers to stop and rethread their sewing machines time and again. "To save 15 cents per spool they were losing $150 per hour rethreading the stuff," he explained. Better thread was purchased and the problem disappeared.

Deming argued pointedly that statistical control is the only sensible means of quality control. "All these companies are running advertisements on the TV about their rigorous inspectors," he snorted. "Inspection is too late, don't you see? It's so obvious. By the time the product gets to an inspector, the quality, either good or bad, is already in. Do you want to burn the toast and scrape it, burn the toast and scrape it—or do you want to make the toast right before it gets to your inspectors?"

The right way to make toast, under the Deming rules, would involve careful observation of every element of the process—the quality of the bread in the market, the level of light in the kitchen, the wiring and the timer in the toaster, the methods of inserting the bread into the slot and of taking the finished toast out. If all these variables were tracked on statistical control charts, a manager could see just what had happened when any piece was burned. The problem could be corrected quickly. The control process would be more efficient and better for workers' morale than inspecting and scraping every piece of toast.

Moreover—and this is the Deming route to enhanced productivity—statistical control is cheaper. It will always cost less to make one piece of toast right than to burn one piece, inspect it, reject it, and then make another piece the right way. "The total cost to produce and dispose of a defective item exceeds the cost to produce a good one," the professor's text says. The time and money previously spent to inspect and scrape burnt toast can be directed to a more productive goal—making more toast—once a predictable level of toast-making quality is achieved.

As the challenge from Japan and other foreign competitors became acute, late in the twentieth century, American industry learned the hard way that the Deming rules should be heeded. In the 1930s, though, things were sharply different; when W. E. Deming talked, nobody listened. There was a brief spurt of interest in statistical quality control at the start of World War II, when quality of production became a matter of national survival. Engineers from munitions plants were brought in to learn from Professor Deming, and the War Department gave courses in factories. "Brilliant applications burned, sputtered, fizzled, and died out," Deming wrote later in his typically acerbic fashion. "Quality control departments sprouted. They plotted charts, looked at them, and filed them. They took quality control away from everybody else, which was of course entirely wrong, as quality control is everyone's job."

Deming was not a man to give up easily, but the wartime experience led him to focus on other applications of statistics, including demographics. In the global reconstruction at the end of the war, he became a sort of roving ambassador of demographics. He was an official observer at the Greek elections in 1946, and from there went to Delhi to advise on the Indian census. Eventually he was summoned to Japan by General Douglas MacArthur during the postwar American occupation to assist in various population and housing studies. And there, at long last, Deming found an audience that cared about quality control.

A Japanese statistician, Dr. E. E. Nishibori, who was familiar with the "Theory of Errors," discovered more or less by accident that the author of that insightful paper was in Japan. Nishibori tracked Deming down: would the American be interested in speaking at a quality control workshop for the Union of Japanese Science and Engineering? He would. He did. Another workshop followed, and another. Someone else arranged for Deming to speak in Tokyo at the Industry Club of Japan, a business round-table whose membership included senior management of every major Japanese manufacturing concern. In that speech, on July 26, 1950, Deming declared, to the astonishment of his Japanese audience, that Japanese quality would soon be the best in the world.

"I predicted," Deming recalled with great relish three decades later, "that Japanese manufacturers would come to dominate world markets and have their competitors crying for protection. At that time I was the only person in the world who believed that. . . . But it was so simple. You could see that this society was receptive to the ideas that are necessary for quality and productivity."

That speech turned out to be decisive. "Once you convince senior management that this will make a difference, the hardest job is done," Deming always said, and Japan was Exhibit A for him. The Japanese were so convinced by what they heard that they invited Deming back to Tokyo year after year long after the occupation had ended to spread his gospel. The Union of Japanese Science and Engineering distributed millions of copies of his books and pamphlets. Tens of thousands of Japanese factory workers, engineers, and executives studied his teachings at regular classes (they still do so today). These diligent students put his ideas into practice, with consequences that shook the world.

When Deming arrived in the shattered remnants of postwar Tokyo, Japanese goods were everywhere considered cheap, shoddy products made from cheap, shoddy materials. By the 1970s, the once-proud American automobile industry bought television advertising to boast that its products were made the Japanese way. The label "Made in Japan" came to represent the world standard of quality for products ranging from steel, automobiles, and heavy machinery to cameras, scientific instruments, and consumer electronic gear.

There is no single explanation for Japan's postwar economic miracle, but the Japanese, at least, have given a good deal of the credit to their American teacher. Deming was one of the few foreigners ever to receive the Second Order of the Sacred Treasure, Japan's premier imperial honor. His name and profile adorn the Demingu Shō (the Deming Prize), an annual industrial award carrying the stature of the American Pulitzer Prizes, which is awarded, Oscar style, in an annual presentation telecast live to the entire nation. The name "Demingu" has become virtually a household word.

And then, fifty years late, the name "Deming" came to have meaning in other countries as well—even in the United States. The Japanese and other competitors led American management to pay tardy heed to the native-born Cassandra; if it were not so, Deming would not have had to squeeze two workdays into one. Sometime between his seventy-fifth and eightieth birthdays, Deming emerged as the guru of what was called the Third Wave of industrialization, following the First Wave at the start of the Industrial Revolution and the Second Wave, which swept in with techniques of mass production. The Third Wave, suited to a world with higher standards and a clearer recognition of limits, focuses on the most efficient possible uses of resources to produce goods of predictable reliability and quality.

Dr. Deming loved to remind people that he saw the new wave coming long before his countrymen caught on. "The way to compete in an international economy is to promote efficiency, to produce better quality than the other people are producing," he would say. "It's so simple. It's so obvious."

The point was somewhat less than obvious to the American semiconductor industry until the startling Japanese success in memory chips brought the message home quite clearly. The Americans took it to heart. The semiconductor firms made quality control—not just inspection but genuine control—a high-priority challenge. It made a difference. In an interview two years after he released his bombshell, Richard Anderson, the Hewlett-Packard executive, said his firm had found marked improvement in the quality of memory chips coming from American suppliers; in fact, he said, the Americans had matched the Japanese on quality standards. This change helped U.S. firms take back a small share of the world market for 64K random-access memory chips from the Japanese.

"U.S. manufacturers, until the advent of Japanese competition over quality, had made a tacit decision that fast, volume output with component testing to cover imperfections in the manufacturing process was more important than high quality," reported a study published by Congress in 1982. "The Japanese instead concentrated on perfecting their production process to deliver higher quality devices. As U.S. firms retool and expand capacity, they have apparently been 'tweaking' their production process to deliver higher quality devices. It may well be, then, that the Japanese ability to use quality as a penetration strategy will not carry over to the next round of competition."

But the Americans did not "tweak" their quality standards enough to head off the Japanese challenge. By the mid-1980s, the sun seemed to be setting on Silicon Valley. Japan's semiconductor giants had become the dominant force in the world semiconductor market. Despite the inspiring presence of the two inventors of the integrated circuit in its midst, the United States fell further and further behind in overall market share. The Japanese lead was greatest in the field of memory chips, a product that played precisely to Japanese strengths for dependable, high-quality production of a tested product. But firms like Hitachi, Fujitsu, and Nippon Denki also began to take the market lead in certain logic chips and analog chips—product areas where the United States had always been dominant. For a while, Nippon Denki (NEC) even competed with Intel in the microprocessor business for personal computers, until the threat of a patent infringement lawsuit led the Japanese to drop out of that market.

The annual market share surveys put out by Silicon Valley research firms like DataQuest and VLSI Research, announced in January each year, became a somber event for American chip makers. Year after year, Japan had a clear lead in global market share for semiconductors overall, and in most distinct market segments. Beyond that, Japanese machine-tool makers pulled ahead of the Americans in the market for semiconductor fabricating machines. The lead in fabricating machinery was, in a way, even more disturbing. If Japanese machine-tool makers gave their Japanese customers first call on each new model, then the Japanese chip companies would have another advantage over their U.S. competition.

The Semiconductor Industry Association, once the proud voice of a global leader, now began to sound very much like the steelmakers and carmakers and television makers that had earlier succumbed to Japanese competition. The SIA began badgering the government for help to stave off the Asian invaders.

To deliver its warning that the Japanese were coming, the men of Silicon Valley selected one of their own—a senior statesman of the industry who seemed perfectly suited to the role. The designated spokesman was articulate, intelligent, and immediately impressive. He complained that the Japanese were cheating, using their monopoly in the closed domestic market to fund predatory pricing in foreign markets. "Many of the Japanese practices are practices that we would see as unfair, illegal, or whatever," the spokesman told a congressional committee. ". . . It may be that we can hope the Japanese will play by the rules of our game, but I don't see any motivation for them to do so whatsoever, since they perceive that they are winning using the current strategy. And, indeed, they may win."

The spokesman who offered this mournful plaint presented the same plea for government help week after week at hearings, seminars, conventions, and press conferences all over the country. He performed the job as if he had been doing it all his life, but he had not. The spokesman had been a physicist, then an inventor, then a corporate manager and a venture capitalist—and had performed all those jobs with equal facility. The spokesman was Robert N. Noyce.

In the first edition of this book, this chapter came to an end at that unhappy juncture for the American semiconductor industry. Bob Noyce's emergence as spokesman for the industry came at a time of desperation among semiconductor firms. There was a feeling, by 1985 or so, that the United States had already lost the race, that American industrial champions like Intel, Motorola, and IBM were bound to be left behind by the Japanese microchip juggernaut.

It was inevitable, in a way, that the U.S. semiconductor community would turn back to its founding father, Noyce, in time of

crisis. As the co-inventor of the microchip, he had technical cre-
dentials that nobody could gainsay. But he also had a stellar
record as a manager, entrepreneur, speaker, and advocate. Almost
everything Noyce ever touched had turned to gold; now he was
being asked to restore the tarnished luster of American competitiveness.

At first, the movers and shakers of the industry headed off in the
same direction as older industries—steel, autos, tires, glass, audio-
video, etc.—that had run up against formidable Japanese competitors. The Semiconductor Industry Association, a trade group
that Noyce had been instrumental in starting, hired some of the
same economists and lobbyists who had worked for other threat-
ened industries. At the urging of these experts, the chip makers
ran to Washington for help. As head of the SIA, Noyce appeared
before Congress and dutifully read from the standard script,
which blamed the problem on the nefarious Asians—"unfair, illegal, or whatever." The industry pressed the federal government
to help, and the Reagan administration responded. In 1986 the
Japanese government signed a "voluntary trade agreement,"
which was actually not voluntary and not an agreement. It was
imposed unilaterally by Washington on threat of closing U.S.
markets to Japanese imports in many areas beyond semiconductors. Japanese companies committed to limit their exports of
memory chips and many kinds of logic chips. It was a quota system, pure and simple. But it gave the U.S. industry some breath-
ing space and a chance to rebuild.

Bob Noyce, however, was never really comfortable with this
protectionist response to the industry's problems, or with his role
in bringing it about. Ever since his days as a junior engineer at
Philco, Noyce had nourished a visceral dislike for businesses that
got into bed with government. In fact, he believed that sweetheart
deals with government were one of the causes of the industry's
malaise: "Price wasn't much of an object in the American space

program, and that may have trained a generation of design engineers to leave cost out of the equation," he wrote, a rather daring admission for an American semiconductor executive to make in 1988.

The bigger problem for Noyce, though, was a nagging sense that his industry was falling into a rut. All his professional life, Noyce had insisted that the worst way to reach a solution was "to approach the problem the way everybody else has." And now the semiconductor industry was going down the same road as every other industry that had run up against stiff Japanese competition. Noyce was simply too honest to argue that running to Washington for trade protection could solve the basic problems troubling Silicon Valley. "The problems are primarily of our own making, so we have to make our own solution," he said. "We need to get back to the leading edge of design, and we need to get better at manufacturing."

Thus it was that Bob Noyce became the chief architect, and the first CEO, of an industry-wide consortium called Sematech (a name carved out of *semi*conductor *tech*nology). To induce the non-California companies like Texas Instruments (Dallas), Motorola (Phoenix), and IBM (Armonk) to sign on, the founders of Sematech agreed that the new organization would not be located in Silicon Valley. And so in 1987, Noyce left his beloved valley and moved to the less verdant precincts of Austin, Texas, to try to save his industry. Somewhat reluctantly, he agreed that Sematech would take government financial help—about $100 million per year—but he insisted that the corporate members must contribute an even larger figure so that Sematech would be a primarily private undertaking.

From the first, Noyce realized that Sematech need not worry much about American skills in semiconductor physics, circuit design, or innovation. The United States already led the world in those fields. Rather, the consortium focused on the more mundane problem of production. The goal, Noyce said, was to see to it that American chip makers could match or exceed the Japanese in speed and quality of manufacturing. The consortium approach, with companies and universities from all over the country working together, was designed to help the relatively small U.S. firms at the forefront of technology—outfits with names like Cyrix, Xilinx, Micron, and Zilog—compete with behemoths like NEC, Hitachi, and Toshiba in microchip manufacture. And the way to do it, Noyce argued, was to make "manufacturing" a dynamic and indeed glamorous field for bright Americans to pursue. The main thing America needed, Noyce told *Fortune* magazine, was "mothers who say proudly, 'My son, the manufacturing engineer.' "

While Noyce hunkered down to this significant but not very sexy task, many U.S. experts sneered at the Sematech concept. The real issue, they argued, was not just one or two troubled industries, but rather a broader problem of American "decline" vis-à-vis the emerging industrial powerhouses of East Asia. In this view, Japan, and its industrial nephews like Taiwan, South Korea, and China, represented the rising sun of a "new economics" while old-fashioned U.S. capitalism was drooping beneath a dark horizon. A distinguished student of Asia, James Fallows, set forth the theory in his treatise *Looking at the Sun*. Fallows opened the book with a long, disturbing chapter relating how the Japanese had come from far behind to trump American industry in the crucial field of semiconductors. It was a tale, Fallows wrote, of "Japanese triumph and American failure."

But Fallows and his fellow chroniclers of American decline had not counted on the formidable talents and energy of Bob Noyce. Traveling the country, hectoring his colleagues, constantly pushing researchers, engineers, and corporate chairmen to work harder and faster, Noyce spent four hard years driving the American industry to make up for lost ground. And he succeeded. Early in 1993 (just as *Looking at the Sun* was hitting the bookstores), the annual figures for global market share in semiconductors were released. For the first time in a dozen years, Japan was number two; the American industry had caught up with and passed its nemesis. The lead wasn't huge—U.S. companies had 44 percent of the global market in 1992, to 43 percent for the Japanese, 8 percent for European makers, and 5 percent for other Asian countries— but it marked a stunning reversal. "It's gone from total disaster to an American triumph," said G. Dan Hutcheson, president of VLSI Research. "This shows that you can turn an industry around with creativity, hard work, and help from the government."

In the following years, Americans cemented their leadership position. By the turn of the twenty-first century, the United States claimed a comfortable 45 percent of global markets, while Japan dropped to about 40 percent, losing some market share in memory chips to its smaller Asian neighbors. In the late 1990s, in a particular tribute to Sematech's achievement, U.S. makers of semiconductor manufacturing equipment once again took the lead in global market share.

While the American industry marched forward into the broad, sunlit uplands of world leadership, though, there was one shadow over the achievement. Bob Noyce, the man who was probably more responsible for the U.S. renaissance than anyone else, was not there to see it.

THE PATRIARCHS

Robert Noyce's metamorphosis from corporate manager to industry spokesman had come about as gradually, and as inevitably, as his earlier transition from inventor to manager. The change was not something he had planned; "I just sort of drifted into it," he said. But then, if Noyce's life had turned out the way he originally planned, he would have spent most of it in a physics lab somewhere, happily indulging his curiosity about the way things work and turning out monographs explicating interesting phenomena of solid-state physics.

In fact, Noyce's career had already begun to move out of the lab by January 1959, when he hit upon the monolithic idea. That was his most important engineering breakthough, and it was also, for all practical purposes, his last. There was something about Noyce—a fundamental confidence matched with a compelling sense of presence—that prompted people to look to him for leadership. The physicists, chemists, and engineers working at Shockley Semiconductor in 1956 were required—it was part of William Shockley's unique management style—to give one another report cards. When the grades were tabulated, Noyce emerged as the consensus choice to be the group's technical director. Accordingly, when the "traitorous eight" left Shockley in 1957 to found

Fairchild Semiconductor, the group turned to Noyce as soon as it became clear that somebody was going to have to act as a manager. His colleague Gordon Moore recalled that "Bob was everybody's choice. He was the natural leader."

In any case, it was a satisfying development, not least because it provided a whole new world of human endeavor to learn about. "Getting into management was just enormously exciting," Noyce recalled a few years later. "Because, first of all, I didn't know a damn thing about it, so that your learning rate goes up very, very rapidly. But secondly, management does become the focal point for all the information in the organization. Well, the guy who has the information has the power. . . . It's a very satisfying thing, particularly coming from a place where you're looking at a narrow field so you don't see the forest for the trees. And suddenly you're sitting in a balloon looking down from branch to branch and . . . for the first time you can see the whole."

More important, it just felt right—it fitted precisely with Noyce's evolving theory of corporate leadership—to have a technologist running a high-tech company. "One of the real problems with American business," he argued, "is this notion that you can be trained in management, in some kind of generic form of management, and then you can manage any operation. But that absolutely does not work in a technical situation. The manager has to have an intuitive gut feel for what ought to be done in a particular situation, and if you don't have the technical background, if you haven't participated personally, you don't have that."

For the most part, the American semiconductor industry has been built and run by technologists, and to Noyce this was a key reason for its explosive success. "They say we were lucky," he explained. "Well, you can say you were lucky being up at bat in the World Series with two out in the ninth inning in the final game of the series and you happen to hit the home run. But the real point is that you have to be at bat at that time. Basically, you have to be in there participating and have the gut feel for what can be done."

Noyce's first management position at Fairchild Semiconductor was director of research and development, a position considerably more important than it might sound because, for the first year or so, the firm was doing nothing but research and development. As the research developed into important products—the planar process and the integrated circuit—the upstart semiconductor division became one of the leading profit centers in the whole of Fairchild Camera and Instrument Corporation.

Fairchild's directors back east rewarded Noyce in the traditional ways: he was named general manager of the division, then vice president, which was basically the same job with a fancier title, then group vice president, which was still the same job but even fancier. His salary began to climb toward the exalted levels reserved for athletes, entertainers, and the top management of large corporations. But after 1959, when he and his seven cofounders sold their $500 worth of stock to the parent company for a quarter million dollars each, salary was not so important a concern.

And yet, through the mid-1960s, Noyce was becoming more and more uncomfortable. Fairchild's directors wanted to run the new profit center their way, and this was something quite removed from what Bob Noyce and Gordon Moore had in mind. The two technologists were, as Noyce put it, "comfortable with risk," be it technical or financial. The generic managers back east at the home office were comfortable only with security, with the safe business play. Looking back later, Noyce could see that a fissure was inevitable. A bungled turnover among top management back east in 1967 left the California group shaking their heads in disgust. That paved the way for the final break, in 1968, when Noyce and Moore decided—in sharp contrast to contemporary corporate wisdom—that money could be made by building computer memory circuits on a semiconductor chip. With the help of Silicon Valley's leading venture capitalist, Arthur Rock, the pair left Fairchild and founded Intel Corporation. They arranged things so that technologists would be in control. The firm's president was Robert Noyce; its chairman was Gordon Moore. Intel was to be governed, Noyce explained, not by market surveys and financial analyses, but rather by "intuitive gut feel."

Noyce related all this one day in his office at Intel's headquarters, a sprawling rectangle of nondescript architecture alongside the Central Expressway in Santa Clara, just shouting distance from an enormous amusement park named Great America. To call Noyce's office an office, though, is to stretch the language. The better term would be "work space." The founder's work space, like the spaces allotted to everyone else at Intel, was really just a small cubicle bounded by four white metal partitions. Intel did not, and to this day does not, provide huge executive offices. Noyce's parking space in the lot outside was whatever space happened to be free when he arrived at work. Intel does not provide reserved parking for the brass. If Noyce arrived at the office later than 8:00 A.M., the official start of the workday, his name went on the late list like all other dawdlers. Intel does not permit exceptions for those at the top. While junior executives of neighboring companies far less successful than Intel routinely rode their limousines to San Francisco for lavish expense-account lunches at exorbitant bistros, Bob Noyce's routine lunchroom was the Intel cafeteria. "The potato salad's pretty good," he told me over his tray.

All of which was a direct reflection of the democratic, meritocratic, and studiously nonhierarchical management style that Noyce, Moore, and another Fairchild refugee, Andrew Grove, custom-designed for their company. They wanted to nurture at Intel a feeling of "nobody here but us engineers"—a sense that the company was just a bunch of technologists working together to solve technical problems and keep two hops ahead of the competition. The thrust was to give each scientist, engineer, and mathematician on the staff the ability—and the responsibility—to work at the leading edge of the industry. This management philosophy, laid out in Andrew Grove's book *Only the Paranoid Survive*, was specifically designed for a company full of bright, curious, and driven superachievers—in short, for a company full of Noyces.

It was also a philosophy that worked. "Intel became Silicon Valley's technology flagship," *The Wall Street Journal* reported on the firm's tenth anniversary, thanks to its pioneering development of semiconductor memory and the microprocessor. Intel struggled through downturns with the rest of the industry, but overall its corporate history has been a story of stupendous growth. A little less than fifteen years after it opened for business with no product and a few dozen employees, Intel became a billion-dollar company. In the year 2000, its sales topped $32 billion and there were 85,000 people on the payroll in four dozen countries around the world.

Intel's corporate success was accompanied by extraordinary financial success for its founders. In 1969, at the age of forty-one, Noyce had become reasonably well-to-do with the quarter-million-dollar profit on his Fairchild stock. After the birth and rapid growth of Intel, though, he moved into the category of the seriously rich. On the day in 1971 when Intel's stock went public—at an opening price of $23.50 per share—Noyce's net worth went into the eight-digit range.

When a man whose self-description included the phrase "comfortable with risk" finds himself dripping with money, it is almost axiomatic that he is going to risk some of it on financial ventures. Robert Noyce, who had gone pleading to venture capitalists when he started Intel, now became one of the more active venture capitalists in the high-tech sector. His portfolio was studded with names like "diasonics" and "monoclonal antibodies." Naturally, he put a good part of his money into the semiconductor and computer industries. But even when investing in the fields he pioneered, he cheerfully admitted that his investment decisions have not always been the wisest. His wife asked him once about a start-up company down the road in Cupertino that was looking for investors. Noyce studied the prospects and warned her to stay away; no future there. The "no-future" company was Apple Computer, which went on to become the very symbol of financial success in the personal computer industry. (Fortunately, Mrs. Noyce had the good sense to ignore her husband's advice.) Bob himself put some money into a seemingly solid outfit called Osborne Computer. It had a dynamic product idea—the world's first portable personal computer. Unfortunately, the Osborne product was so heavy and cumbersome that it came to be known not as a "portable" but as a "luggable." The firm lurched into bankruptcy, taking Noyce's money with it.

There were, however, more Intels than Osbornes in Noyce's investment record. By the time he left Intel for Austin in 1987, the minister's son from Denmark, Iowa, was counting his assets in the billions of dollars. (His alma mater, Grinnell, was also in the chips. Noyce had arranged for the school to buy a hunk of stock at Intel's initial public offering, an investment that made Grinnell one of the richest small colleges in the world.) Noyce used the money to pursue his unbounded appetite for new experiences. He flew his own jet to skiing vacations at his condominium in Aspen. He flew his own seaplane to boating vacations on the lakes of northern California. Comfortable with risk at play as well as work, he took up gliding, hang gliding, paragliding, scuba diving, and assorted other sports, just for the hell of it. At Aspen, he arranged his day so that he worked in the afternoon and spent the mornings racing down Ajax, the toughest slope on the mountain. I asked him once to name his favorite run at Aspen, and he didn't hesitate a minute:

"Face of Bell," he said. When I trekked over to see this run, it turned out to be a precipitous cliff marked with double black diamonds and a sign in red letters: "Caution: Experts Only." He bought ever fancier cars and drove them ever faster. He took up fine wines, not surprisingly becoming an expert in enology. As long as he lived in Silicon Valley, he stayed at the comfortable but hardly fancy house in Los Gatos where he and his first wife (he was divorced in 1974 and remarried a year later) raised their four children. But the house was now surrounded by increasingly lavish gardens—in his mid-forties, Noyce became intensely interested in gardening—accoutered with tennis court and swimming pool. When he moved to Austin, he put in a garden and pool at his new home there, and spent what little time he had outside of work developing an expertise in native Texas flora. "He had one helluva life," wrote Michael Malone, who worked in the semiconductor industry when Noyce was its most famous citizen. "In Silicon Valley, all of us wanted to work for Dave Packard [of Hewlett-Packard], but all of us wanted to be Bob Noyce."

The combination of Noyce's vigorous athletic life, his far-flung personal interests, and his management work at Intel would have been sufficient to fill the days of most men. For Noyce, though, his late fifties found him looking beyond the low partition in his cubicle at Intel. Among much else, he started thinking about the future not only of his own company but also of the entire semiconductor industry—and of American industry as a whole. He was among the leaders of a group of Silicon Valley executives who got together to form the Semiconductor Industry Association, and was, of course, a charter member of the board. When journalists came to California to look into the chip business, Noyce's office became a regular stop. He was, in many ways, the perfect spokesman—a thoughtful, articulate founding father who could discuss both precise technical questions and broad issues of industrial policy.

By 1978, Noyce was spending as much time speaking at conferences, seminars, and congressional hearings as he was at Intel. He stepped aside from day-to-day management—turning the firm over to his old friend and sounding board, Gordon Moore—so that he could devote more time to industry-wide concerns. He was the elder statesman of Silicon Valley, the "scientist-cum-charmer," as a colleague put it—the official voice of his industry. He became chairman of the Semiconductor Industry Association and the leader of its battle to stave off the challenge from Japan. It was his impatience with the protectionist approach to the Japanese challenge that got him thinking about other ways to revitalize the industry that he had started. That led to the formation of Sematech. Noyce was one of the key voices calling for an industry-wide consortium, and when fourteen major companies agreed to put money into Sematech, Noyce served on the committee set up to find a CEO for the joint operation. The usual headhunter firms were contacted, and several people with stellar résumés were interviewed for the job. At the final meeting of the selection committee, though, the other members staged a mini-revolt. They threw out the list of candidates and turned to the man who had been the obvious choice from the beginning: Bob Noyce.

At Sematech, Noyce turned himself into an expert on the production end of the chip business—the incredibly precise photolithographic machinery that "prints" wires less than a millionth of an inch thick onto razor-thin wafers of silicon. He took excursions into new terrain that might well have been marked "Caution: Experts Only." Now he was dealing with metallurgy, laser projection, submicron optics, and other complex areas that Noyce had generally left to others at Intel. His goal, ambitious but at least clear, was to see to it that American-made semiconductor fabrication machinery was the most advanced in the world. His bet was that if the United States could regain its lead in the manufacturing end of the microchip trade, it would once again become the world's leading supplier. Then, proud technologists from Silicon Valley would no longer have to crawl to Congress begging for protection from foreign competition.

The goal was still some ways off when I saw Bob Noyce in the spring of 1990. He was friendly but intense, as usual, smoking like mad and talking at his normal super-fast pace. But he was also confident. "This was a bigger job than even I thought it was going to be," he said. "But I think we're going to do it. We've identified the right solution, I think, and we definitely have the talent and the will to achieve it. The Japanese are excellent manufacturers and serious competitors. But they have some problems of their own. So I think we're going to get there."

When the U.S. semiconductor industry did get there, though, just two years later, Bob Noyce was not around to see it. On a Sunday morning in June 1990, he got up early at the house in Austin and dove into the pool for his daily laps. Almost immediately, his heart stopped. They fished him out of the water and got him to a hospital, but within an hour he was dead. He was a son of the prairie who had gone enormously far in his sixty-two years. But he didn't get to arrive at his final professional destination.

Among those who mourned the passing of this giant of American engineering was Jack St. Clair Kilby. Jack told the reporters who called that he was filled with admiration for Noyce's accomplishments, both as an inventor and as the official voice of the industry. Then he sat back in his cluttered office at the northern edge of Dallas, quite happy to be the official voice of no one but himself. He has remained there pretty much ever since, perfectly willing to talk to the occasional reporter or historian who stops in, but for the most part tacitly and creatively engaged in the business he has always liked best: inventing.

For a while, in the years after he finished the Pocketronic calculator, Jack Kilby, too, seemed to be moving more or less inevitably into management. Eager to let him know that his groundbreaking work was appreciated, Texas Instruments rewarded Jack with a steady flow of raises, bonuses, stock options, and promotions. Promotion, for a man whose job was inventing, involved moving up to a management position supervising other inventors; eventually he became the number two man in the hierarchy at TI's research and development lab, and the top job there was easily within reach. By 1970, when he went to the White House to receive the National Medal of Science, he was TI's most respected engineer; he seemed assured of more bonuses, more raises, and more promotions as long as he stayed at the company.

And then, in November 1970, he left.

Over the years, Jack had been thinking a great deal about the work he loved most: the demanding, creative, and ultimately rewarding job of inventing. It became more and more apparent to him that real creativity, artistic or technical, demanded real freedom—the kind of freedom that did not mesh with bureaucracies, governmental or corporate. The whole organizational structure of the corporation was beginning to chafe. "There is a basic incompatibility of the inventor and the large corporation," Jack wrote in a lecture on the subject. "Large companies have well-developed planning mechanisms which need to know at the beginning of a new project how much it will cost, how long it will take, and above all, what it's going to do. None of these answers may be apparent to the inventor." If Edison had worked for a big corporation, Kilby went on, there might have been no Edison light bulb, because the company's goals might not have matched the inventor's. As he contemplated the work of Edison and Bell and other freelance inventors—men who had defined the problems on their own and worked out their own solutions—Jack began to perceive that the grass might indeed be greener on the freelance side of the fence. He had, as a matter of fact, enjoyed considerable freedom at Texas Instruments, but even there bureaucracy was beginning to infringe. His goal had always been to find "nonobvious" solutions to important problems; but anything nonobvious was anathema to corporate planners and accountants.

From his voluminous reading, Jack had put together a pair of lists comparing major inventions produced by large corporations to those made by individuals. The corporate inventors' list included, among other things, Scotch tape, television, and nylon—not to mention the integrated circuit. The individual inventor was credited with, among other things, air-conditioning, penicillin, xerography, and the zipper. Looking back on it today, Kilby is at a loss to explain any significant difference between the two lists. At the time, though, as he perused the two lists, they somehow seemed to prove a point. Today he realizes that the point had already been largely settled in his own mind. Freelance was the way to go.

And so he packed up his books, his papers, and his favorite old Log-Log Decitrig slide rule and moved out to a place of his own in a low-rise office building off the LBJ Freeway in North Dallas, about two miles from TI. There he could work at his own pace on problems of his own choosing. "There's a certain amount of satisfaction," he said, "in setting your own goals, in being free to do what you decide is important, and not pursue somebody else's schedule. The freedom, that's what interests me about this."

He admitted readily that the choice of freedom was not a particularly wise one in financial terms. "No, it was pretty damn close to stupid. The economic rewards have been pretty marginal."

The economic rewards for Jack Kilby have never approached the vast wealth accumulated by Robert Noyce, a fact that Kilby accepted philosophically. "Basically, engineers are hired by companies to do that kind of work," he said. "I don't get five percent of the value of everything that is ever sold, or anything of that sort . . . but I was rather well rewarded by TI for my work on the integrated circuit." His rewards at TI put his yearly salary safely into the six-figure range, an income that permitted Jack and Barbara and their two daughters to live in comfortable upper-middle-class style. It also permitted them to save enough to maintain their way of life during the lean years when Jack first set out on his own. Jack would eventually draw some income for consulting work at Texas Instruments and as a faculty member at Texas A&M's Institute of Solid-State Electronics, a part-time appointment he held from 1978 to 1984. Over time, as well, considerable money started pouring in from prizes Jack was given for launching the microelectronic revolution; he got $400,000 for winning Japan's Kyoto Prize in 1993, and just under a half million for the Nobel Prize in 2000. But much of that money he gave away. Being rich never mattered much to Jack Kilby.

As an independent inventor, Jack has received about a dozen patents, and they reflect the considerably wider scope of his ideas since he went out on his own. At his wife's suggestion, Jack started working on an "electronic intercept" device that keeps your telephone from ringing unless the call is one you want to take. Three years of tinkering resulted in Patent No. 3,955,354, "System for disabling incoming telephone calls." The gadget works perfectly but so far has been a dud in the market. Then there was Patent No. 4,001,947, "Teaching system," a small calculatorlike device that talks to a student as it teaches math, spelling, and other subjects. That idea, too, worked fine, but in this case Kilby was scooped by a competing product known as Speak 'n Spell—produced by Texas Instruments. The Kilby "Electronic check writer," Patent No. 3,920,979, was licensed by a Japanese consumer electronics firm but has yet to earn its first dime, a description that also holds true for Patent No. 3,944,724, "Paging system with selectively actuable pocket printers."

For the first few years or so of his freelance career, Kilby reveled in his new freedom, moving from one idea to the next as the spirit took him. As always, he remained an engineer, not a scientist. His goal was to solve problems, and he now had the opportunity to choose which problems he would work on. "There are a large number of real needs which the inventor can address," he said in his lecture on inventing. "The individual is free to choose a need that he thinks he may be able to satisfy. . . . *The definition of the problem becomes a major part of the innovation*" (Kilby's emphasis).

For several years, he worked on a problem of major importance: the nation's energy supply. After the oil shocks of the 1970s, a consensus grew that there was a fundamental need for some other source of power—a source that was available to everyone on earth, a source that was unlimited, a source that didn't require huge smokestacks emitting greenhouse gases. In fact, this power source already existed—the sun. The problem was how to turn sunshine into electricity. It was a good problem for Jack, because, like the tyranny of numbers, it is one that apparently can be solved with semiconductors, particularly silicon. The basic idea here, known as the photovoltaic effect, was one of the earliest discoveries of semiconductor physics. Physicists have known for more than a century that if you shine light on a strip of semiconductor material, such as silicon, electrons will start to move from one end of the strip to the other. The flow of electrons, as J. J. Thomson explained, is an electric current. Jack defined the problem as building a solar cell of semiconductor material that would fit in an average house and generate enough power to meet the daily needs of that house. It was not an easy problem. "I can't say exactly that nothing came of that work," Jack said. "But not a hell of a lot came of it."

Gradually, Kilby found himself drawn back toward the corporate world. He felt a loyalty to Texas Instruments, and when TI asked him to consult on this problem or that one, Jack was happy to oblige. He joined the board of directors of two smaller semiconductor companies, and consulted on projects at various other firms and research centers. And he kept tinkering away at his inventions.

For a man whose notion of heaven is to seize an interesting problem and solve it, Jack Kilby's lot has been a happy one. Today, in his late seventies, he leads a quiet, thoughtful, and satisfying life. His wife died in the early 1980s, shortly after the couple's thirty-third anniversary. But his sister Jane, as talkative as Jack is quiet, lives in Dallas, not far away, and watches after practical matters. She's the one who can find Jack's glasses and hearing aid when they disappear. He sees his two daughters, Ann and Janet, and five granddaughters often. One of the great pleasures of his life has been taking his family with him around the world to award ceremonies. "You should have seen the look in the principal's eye," said Ann Kilby's daughter, Katrina, "when I told her I was going to take a week off school because my granddad won the Nobel Prize."

Most Fridays, Jack digs out his old TI employee badge and heads over to the R&D lab at Texas Instruments—now located in a sparkling new facility known as Kilby Center—to hang out with his fellow engineers. Texas Instruments' press releases refer to this as a "regular consultative role." Jack described it a little differently: "I kinda wander around and talk to people."

And he still spends a lot of time thinking. "Jack is a thinker," said his fellow retiree Willis Adcock, the sprightly engineer who brought Kilby to Dallas a quarter century ago to tackle the tyranny of numbers. "I would say Jack's got a good creative sense, he's got that, but the other thing that I liked when I hired him is that he is a persister. He just thinks a problem all the way through, works it through, and he doesn't stop until he's got it worked out. And you know, you can see the results."

You can see the results anywhere, because the Kilby-Noyce microchip has become an ever-present reality, and necessity, of modern life. Just as it proved impractical back in 1931 to turn off all the world's light bulbs for two minutes as a memorial to Thomas Edison, it would be unthinkable for anybody now to suggest shutting down all the world's integrated circuits in tribute to its inventors. For one thing, you couldn't find them all. Microchips by the billions are at work in homes, cars, pockets, artificial organs, offices, hospitals, and factories around the world. The monolithic idea has spawned thousands of companies and tens of thousands of new products.

It has also spawned a veritable flood of awards, plaques, prizes, and citations for Jack Kilby and Robert Noyce.

The professional and technical awards became so common, in fact, that by the twenty-fifth anniversary of their invention, neither Noyce nor Kilby paid much attention when word of a new one came in the mail. One day in 1982, though, Jack Kilby received an honor that really mattered to him, because it was proof he had succeeded at his chosen trade. He was inducted into the National Inventors Hall of Fame, an august group of just five dozen people at the time—among them Edison, Bell, Ford, Shockley, and the Wright brothers.

On a sunny winter Sunday, a group of people gathered in the lobby of the Patent Office, just across the Potomac from the Washington Monument, for the Hall of Fame induction ceremony. Of the five inventors honored that year, only two—Kilby and Max Tishler, who started the vitamin industry in 1941 by synthesizing vitamin B_2—were still alive. Both were present. When the secretary of commerce called out Tishler's name, the aging chemist stood up and gave a long speech about how he got his idea and what it had meant. Then it was Jack Kilby's turn. He stood up for the briefest moment, looked around shyly at the audience, and quietly said, "Thank you."

A scattering of newspapers around the country ran a brief story on that ceremony and on Jack Kilby's role as the patriarch of digital circuitry. A year later, when Robert Noyce was inducted into the same Hall of Fame for his part in the creation of the chip, a few papers again devoted a few inches of space to the subject. Thereafter, Noyce used to show up in newspapers or business magazines in his role as the spokesman for Silicon Valley. But neither Noyce nor Kilby ever received enough attention in the press to make their names familiar to any more than a minute fraction of their countrymen. Indeed, nearly half a century after they came up with an idea that launched the microelectronics revolution, both Robert Noyce and Jack Kilby remain cloaked in obscurity.

It's a sign of the times. A few generations ago, men of this ilk—men like Edison and Bell, Ford and Goodyear, whose inventions touched every life and spawned giant industries—were accorded enormous prominence. Although such things were not surveyed as carefully then as they are now, it seems fairly safe to assert that Thomas Edison was the best-known man in the country, and probably on earth, within ten years after he perfected the light bulb. (The best-known person on earth in 1890 was almost certainly Queen Victoria, whose portrait hung on walls throughout the world's vastest empire; but Edison was the best-known man.) The Wizard of Menlo Park, the "Napoleon of Science," he still ranked as the "Most Admired American" in a *New York Times* survey taken in 1922, when he was seventy-five years old and long finished with productive work. Alexander Graham Bell was a household name on the basis of his invention long before the nationwide Bell System was in place. Henry Ford and his Tin Lizzie became the stuff of myth, instantly recognized around the world as the symbols of the automobile age.

The chip has changed the world as decisively as did the telephone and the automobile. And unlike many modern inventions, we know exactly who gave it to us. But in the microelectronic age, Jack Kilby and Bob Noyce symbolize, if anything, only the modern lack of interest in the humans behind the machines. Barely one American in ten thousand could name the two countrymen who invented the integrated circuit and launched the digital revolution. They are not the stuff of which heroes are made in contemporary American society.

Is it because people really believe that computers contain "electronic brains"—and thus don't care to know about the human brainpower that made these mechanisms possible? Is it because we have swallowed the Orwellian notion that digital technology is a brutalizing, tyrannical force—and thus we don't want to honor, or even know, the men who made it? Is it because we have grown so accustomed to new ideas coming out of huge corporate and governmental enterprises that we no longer recognize individual invention? Is it because the media that purvey fame and recognition among our contemporaries—*People, Oprah, Larry King, Good Morning America,* and the like—don't trust their audiences to appreciate genuine intellectual accomplishment? Or maybe it's because wealth matters more than achievement when it comes to choosing the people our society will look up to. Bill Gates was an accomplished and innovative programmer who launched the global industry of personal computer software; but who had ever heard of him until he showed up in *People* magazine as "the richest man in the world"? A *Time* magazine story about Bob Noyce shortly before his death focused on his investment earnings and described him as a "financial genie"—a classic case of missing the real point.

The list of the "most admired" in today's world—a list assembled annually by the sophisticated surveying apparatus of the Gallup poll—suggests that the current vogue in admiration among Americans runs heavily to political figures, with an occasional clergyman or entertainer thrown in. For decades George Gallup and his organization have been asking Americans to name the two people they admire most. The answers vary little from year to year. The pope and Billy Graham are often on the list. Bob Hope and Walter Cronkite show up occasionally. The other names tend to be drawn from government; Ronald Reagan, Bill Clinton, and Colin Powell are among the hardy perennials in this garden. As the Gallup organization points out in a caveat accompanying its survey, the poll "tends to favor those who are currently in the news." It's hardly surprising, then, that men and women engaged in science and engineering tend to be left out, for such people are generally not treated as news—unless they become avid self-promoters (as Edison and Ford were) or unless, like William Shockley, they set aside their technical work and begin proselytizing for political causes.

And so, in an era when everybody is supposed to be famous for fifteen minutes, Jack Kilby and Bob Noyce have never come into their allotted quarter hour. There have been occasional stories about them in newspapers and magazines, particularly in the local media of Dallas and Silicon Valley. In Dallas, as a matter of fact, Jack Kilby is almost a minor-league celebrity, largely because of genuine pride in the hometown boy who did well. The local media call him the "Texas Edison." There's an official Texas Historical Marker on the site of the lab where Jack first wrote down the monolithic idea. Texas Instruments is particularly proud of the eminent inventor in its ranks, and that has increased Jack's stature in Dallas. When the company opened its new Kilby Center in 1997, it hung a massive canvas poster on the wall of the building, visible to every commuter whizzing by on I-635: "The Chip That Jack Built Changed the World." When a group of Dallas citizens created an annual award for people who contribute to human happiness through science—the laureates include physicians, chemists, botanists, etc.—they gave their prize the most appropriate name they could think of: the Kilby Awards.

Next to Dallas, the one place where the name "Jack Kilby" is fairly broadly recognized is Japan. The country that honors W. Edwards Deming also honors the inventor of the microchip, a product that helped lift the Japanese to global industrial prominence in the 1980s. This is partly because Jack's name was a hot news item in Japan for almost two decades while Texas Instruments was battling Japanese semiconductor firms for patent license fees. The original TI patent for the integrated circuit is known in Japan as the "Ki-ru-bee tokkyoken"—that is, the "Kilby patent"—and thus the name was for years a common term on Japanese front pages. Beyond that, Jack Kilby is exactly the kind of person that Japan tends to admire. The founder of Sony, Akio Morita, another technologist who achieved heroic stature in Japan, said that the real key to his island country's economic success was that "we are a society that honors engineers." So it is natural that the Japanese would honor an engineer who changed everything.

Accordingly, Jack has traveled regularly from Texas to Tokyo, where the media literally line up for interviews with him. When Texas Instruments opened a big new engineering center in the "science city" of Tsukuba, Japan, the company, of course, sent its best-known engineer to do the honors. The ribbon-cutting ceremony was performed in characteristic leave-nothing-to-chance Japanese style. Jack and a host of Japanese officials were each given white gloves and golden scissors festooned with white, red, and purple streamers. They were directed to their assigned places at the entrance of the new building, and an announcer instructed them: "Hold the ribbon in your right hand, hold the scissors in your left, bring the ribbon up to the blade, pause ten seconds for the cameras, and now—CUT!" With the ribbon successfully cut, the announcer offered a breathless replay of what had just happened. "We are deeply honored," he ended, "that an engineer who has won the esteem of all Japan could join us today to open our humble center."

Back home, Jack could not quite claim the esteem of all America. But there were moments. Jack's first integrated circuit was presented to the Smithsonian Institution, where—unlike Keuffel

& Esser's last slide rule—it is on permanent display in the National Museum of American History. The U.S. Postal Service issued a 33-cent stamp honoring the invention (but did not include the names of the two Americans who invented it). And shortly after Jack Kilby joined Ford and Edison in the Inventors Hall of Fame, Diane Sawyer flew to Dallas to interview him for the CBS Morning News. The segment lasted about three hundred seconds, with Sawyer tossing out peppy questions and Jack responding in his slow, laconic way.

"I mean, if you have to think of one thing that kept the United States at the forefront of technology," Sawyer said, "it was really your invention." Kilby paused, stewing it over. "Well, I hadn't thought of it in those terms," he said quietly. "Have you made money from this invention?" Sawyer asked. "Some, yeah," Kilby replied. Things were just starting to get interesting when Sawyer got a signal from the director: time to move on. She turned quickly to the camera and said, "Coming up in a moment, Dr. Jerry Brodie on how to handle the death of a pet." Jack Kilby's moment in the sun was over.

Of course, if Jack had remained in the sun much longer, he probably would have been running toward the shade. The man is so down-to-earth, so genuinely modest, that he seems uncomfortable when people get wound up about his inventions. Far from encouraging interest in his achievements, Jack tends to play them down. When Texas Instruments opened its Kilby Center, the company magazine interviewed, or at least tried to interview, the center's namesake about the monolithic idea and its consequences. "Did you have any idea that you were going to have such a profound effect on everybody's daily life?" Jack was asked. "Well," he replied, "I don't know that I get credit for their profound effect. . . . What you see today is the work of probably tens of thousands of the world's best engineers, all concentrating on improving the product, reducing the cost, things of that sort." The interviewer persisted: "Is it nice to know that you made one of the most major contributions in making Texas Instruments what it is today?" Jack was having none of it: "Well, there's a lot of water under the dam since that time," he said, "and it's hard to take any direct credit for the TI of today."

One thing that Kilby particularly insists is that he not be treated as somebody special. When a school board member back in Kansas proposed changing the name of Great Bend High School to Jack S. Kilby High, the school's most distinguished alumnus immediately scotched the idea. "When they play football against Dodge City, they don't want to be 'Kilby,'" Jack said. "They've got to be 'Great Bend.' Anyway, the whole thing would be a lot of trouble. I'm not worth the fuss."

In the fall of the year 2000, though, Jack Kilby found himself at the center of an enormous fuss. In the predawn hours of October 10, reporters in Europe started placing frantic calls to "J. Kilby" in Dallas. Jack does, in fact, have a listed telephone number, but he's in the directory as J. S. Kilby. The J. Kilby who received those calls was his sister Jane. "As soon as the first fellow said the words 'Nobel Prize,' I knew it was going to be a busy day," Jane said later. She raced over to her brother's house and found a cluster of reporters on the porch, banging fruitlessly on the front door of a man who had removed his hearing aid for the night. Eventually, Jack came to the door in a green robe and expressed gratitude to the Royal Swedish Academy of Sciences. Then he headed back to the kitchen for breakfast.

But it was evident that Jack Kilby's cherished life outside the spotlight was going to change, at least for a while. As the oldest, richest, and most prestigious honor in science, the Nobel Prize is the big enchilada of global awards, the ultimate accolade. The media might have looked the other way when Kilby won the Kyoto

Prize in Advanced Technology or the Holley Medal of the American Society of Mechanical Engineers or the Institute of Electrical and Electronic Engineers Medal of Honor. But the Nobel Prize was not to be ignored.

A dozen years earlier, when it was already clear that the integrated circuit was a development of historic dimensions, I had asked both Jack Kilby and Robert Noyce why their invention hadn't won them a Nobel Prize in something. Both men gave the same answer: The monolithic idea wasn't the type of thing that won Nobel Prizes. "Basically, we're looking at an engineering development," Noyce replied, after noting that he was flattered to be asked. "If you look at the Nobel Prize, it goes to important discoveries in science." Maybe so, I shot back, but the transistor won a Nobel Prize, just nine years after it was invented. The microchip was already thirty years old—and nary a word from Stockholm. "But that's different," Noyce said, in the tones of a patient teacher. "The 1956 prize was not awarded for the invention of the device. It was for the physics of the transistor effect."

Actually, there has been considerable criticism of the Royal Swedish Academy for the excessively esoteric nature of its science prizes, and its failure to recognize the microchip was Exhibit A for the critics. Nobel laureates in physics and chemistry win the award for important discoveries, of course, but often the discoveries are important only to a small clique of specialists in a subset of the field. By honoring the academic and ignoring the practical, the Nobel Committee was arguably slighting the work of people who made genuine contributions—and frustrating the design of Alfred Nobel, who stated explicitly that his award was intended for "those who . . . shall have conferred the greatest benefit on mankind." In the last half of the twentieth century, there was no scientific development that conferred greater benefits than the chip—but there was no Nobel for the chip's inventors. By ignoring, for more than three decades, the inventors of the microchip, the committee denied the prize completely to Robert Noyce. Noyce was dead by the time the Nobel people honored the integrated circuit, and Nobel Prizes are not awarded posthumously. *The Wall Street Journal* spotted in this a plot by European left-wingers to thwart Noyce's legitimate deserts: "Why? Because he was a businessman and because his work had been commercial. The Swedish Academy would never stand for a dirty capitalist, especially a very, very rich one, getting the prize."

For its first award of the new century, however, the Nobel Committee adopted a thoroughly practical stance. The official announcement that went out from the Royal Swedish Academy on October 10 explained that the Nobel Prize in Physics for 2000 recognized the connection between physics and the information age of computers and telecommunications. Accordingly, the physics prize was split. Half would go to Jack S. Kilby "for his part in the invention of the integrated circuit." The other half of the prize money was to be divided between two academic physicists, Zhores I. Alferov of Russia's Ioffe Physico-Technical Institute, and Herbert Kroemer, of the University of California at Santa Barbara, "for developing semiconductor heterostructures used in high-speed and opto-electronics." In essence, Alferov and Kroemer had made valuable improvements on Jack's basic idea. The term "heterostructure" refers to a chip made up of alternating layers of different semiconductors. Alferov and Kroemer had independently determined how heterostructures could be used to generate a laser beam—such as the one in the read/write head of a CD player—and the tiny but powerful amplifier required to make cellular phones work. The academy's announcement conceded that "the integrated circuit is more of a technical invention than a discovery in physics." But that was all right, because "it is evident that it embraces many physical issues . . . [such as] how to produce dense layers that are only a few atoms thick."

As October 10 dawned over Texas, it quickly became obvious

that Jack's brief appearance on the front porch was not going to satisfy the media hordes. Accordingly, later that day, the newly named laureate was ushered into a conference room at Texas Instruments for a formal press conference. Jack first offered a tribute to Robert Noyce: "I'm sorry he's not still alive. If he were, I suspect we'd share this prize." He then answered endless questions about his invention and himself, particularly about his own low-tech lifestyle. "Well, I may be the only person in the room without a cell phone," Kilby conceded. As always, he did his best to deflate the suggestion that he had changed the world, insisting that many other engineers shared the credit. The reporters were smitten by this understated performance from a world-class overachiever. "He looked like everyone's grandfather," observed *The Dallas Morning News*, "not the man who spawned the Information Age."

Two months later, with granddaughters, daughters, son-in-law, and sister in tow, the man who spawned the information age arrived in Stockholm for a glittering round of luncheons, receptions, royal audiences, and the like. "I'm not worth the fuss," Jack said again, as he climbed into yet another Volvo limousine for yet another function in his honor. Still, he appeared to be enjoying himself. Several old friends were on hand to share the occasion. One of the more poignant moments of the week came in the lobby of Stockholm's Grand Hotel, when Jack spotted another acquaintance from years back: Gordon Moore. Bob Noyce's engineering and entrepreneurial colleague, the man who was on the receiving end when Noyce had first enunciated the monolithic idea forty-one years earlier, routinely received invitations to the Nobel Prize ceremony. But now for the first time he had decided to attend. "I did it for my friend," Moore said. "The Nobel Prize for the integrated circuit would have been shared by my colleague Bob Noyce if he were alive. I thought I should come so that Bob would have a presence when his invention was honored."

Other than showing up in white tie and tails for the lavish awards ceremonies—the event is so fancy that even the traffic cops outside wear tuxedos, and the sterling silver laid out for the ensuing banquet is never used for any other function—a Nobel laureate's only unavoidable duty during prize week is to deliver a lecture. Jack Kilby's Nobel lecture in physics took place in a classically Scandinavian lecture hall, all blond wood and sleek modern furniture, on the campus of Stockholm University. Jack was introduced by a Swedish physicist who noted that "Dr. Kilby's" invention had launched the global digital revolution, making possible calculators, computers, digital cameras, pacemakers, the Internet, etc., etc. Naturally, Jack wasn't going to let that go unanswered. "When I hear that kind of thing," he said, "it reminds me of what the beaver told the rabbit as they stood at the base of Hoover Dam: 'No, I didn't build it myself, but it's based on an idea of mine.' " Everybody liked that joke, so Jack quickly added that he had borrowed the story from Charles H. Townes, an American who won the physics prize in 1964.

Kilby then went on, in his deep, slow Kansas drawl, to describe the tyranny of numbers and the various efforts to overcome it in the 1950s. He described the solution he hit upon at TI and explained how Bob Noyce's work shortly thereafter had complemented his own approach. He showed the slide of his original phase-shift oscillator, with the hand-carved chip glued to a glass laboratory slide and the flying wires sticking out every which way. "Had I realized that I would have to look at that thing for 42 years, I would have put a little more effort into its appearance," he said. Jack expressed astonishment at the way the invention had changed electronics: "The development in the last 42 years has been more rapid than in the first 400 years after William Gilbert coined the word 'electricity.' " Improvements in the chip, and particularly the constant cost reduction, had made high-tech wonders available to ordinary people almost everywhere, he went on. "I'm happy to have had even a small part in this process of turning human ingenuity and creativity into practical reality."

On that characteristically humble note, the man who made the microchip strode back to his seat. And, as it turned out, back to obscurity.

For a few weeks, just before that awards ceremony, the Nobel Prize seemed to trigger a new surge of interest in Kilby, in Noyce, and in the broader notion that the two inventors who launched a revolution deserved the recognition and esteem of their country-men. Something about the turning of the century, about America's stature as the world's dominant political, financial, and industrial power at the dawn of a new millennium, led to a na-tional focus on how the United States—which had been, after all, a mere snip of a developing nation one century before—had achieved its exalted status. It was obvious that technologists like Noyce and Kilby were a key part of the answer. And the Nobel Prize in the year 2000 seemed to emphasize the point. Many Americans were proud. Texas Instruments put a huge new poster on that building beside I-635: "The Chip That Jack Built Won the Nobel Prize." The outgoing president, Bill Clinton, invited Jack Kilby to the White House for coffee. The incoming president, George W. Bush, in one of his last official acts as governor of Texas, welcomed Jack Kilby as a charter inductee into the new Texas Science Hall of Fame. Newsmagazines and cable talk shows actually paid some attention to the integrated circuit and the man who had made the first one. The *Topeka Beacon-Journal* named Jack St. Clair Kilby its "Kansan of the Year" for 2000. The front-page article announcing the choice was a textbook example of that old journalistic rule, Find the Local Angle. Harking back to the blizzard of '37, the story began: "A Kansas ice storm more than 60 years ago set the wheels in motion for an invention that would change the world."

But the attention would not last much longer than a Kansas ice storm. By the time the newly honored laureate arrived home from Stockholm, he could stroll through Dallas–Fort Worth Airport without the turn of a single head. Jeff Crilley, a reporter for KDFW-TV, the local Fox affiliate, found a clever way to report on the Dallas engineer who invented the chip. He took a microphone to downtown Dallas and approached passers-by on the street. "Do you have a calculator?" he would ask. "Do you have a cell phone? Do you have a computer at home?" Almost everybody did. "And do you know who Jack Kilby is?" Nobody did. A similar experiment in Washington, D.C., produced the same blank looks. Even in Great Bend, Kansas, of all places, the Nobel Prize came and went with minimal impact on local knowledge about the homegrown Edison. "If I went around town and said the name 'Jack Kilby,'" sighed Jennifer Schartz, editor of the *Great Bend Tribune*, "well, a few people would say, 'Oh, something to do with computers?'"

And there was the central irony. Our media-soaked society, with its insatiable appetite for important, or at least interesting, personalities, has somehow managed to overlook a pair of genuine national heroes—two Americans who had a good idea that has improved the daily lot of the world.

AUTHOR'S NOTE

This book began, some two decades ago, with a disappearing typewriter.

On November 3, 1980—the day before Ronald Reagan was elected president of the United States—I returned to my desk at *The Washington Post* after a year of constant travel covering the presidential campaign. To my distress, my cherished old typewriter had disappeared.

In its place, the *Post* had installed a computer terminal; from now on, I was to write my stories on that. At first, I resented this impostor, as most new computer users did in those early days of the digital revolution. Quite soon, though, I came to realize that this hulking presence on my desk was faster, quieter, easier to use, and far more efficient than the typewriter it had replaced. Within two weeks, I was a devout convert to the new technology.

One day I got mad at something and took a swat at the machine. Red lights blinked; beepers beeped; the screen went dark. In those pioneering days, any company brave enough to install a bunch of computers had to keep a corps of technicians on hand to watch over the machines (and their users). So a technician quickly came over to my broken terminal and opened the cabinet. "We'll have to replace a chip," she said. She pulled out a small black rectangle, maybe half an inch long, with a row of copper legs along each side—a plastic beetle—and dropped it into my palm. "This chip is the heart of the whole thing," she said.

Up to that moment, I had known in some abstract sense that the information age was based on a tiny "microchip," although I had no idea what this term might mean. Suddenly that abstract awareness was translated into a tangible reality in the palm of my hand. I was determined to find out more about this chip and how it worked. Right then and there I dashed out to Reiter's technical bookstore on Pennsylvania Avenue to find a book on silicon chips. The salesman gave me the standard text—*Microelectronics*, by Professor Jacob Millman of Columbia University. This was a legendary book in its field, having nurtured a generation of electrical engineers. For the non-engineer, though, it was tough going; the first paragraph of the first chapter dealt with the impact of a nonuniform concentration gradient on the transport of charges in a crystal. But Professor Millman also recognized the great story that underlay his topic, so he prefaced his text with "A Brief History of Electronics." There I found the following: "In 1958 Kilby conceived the Monolithic Idea, that is, the concept of building an entire circuit out of germanium or silicon. . . . About this same time, Noyce . . . also had the monolithic-circuit idea."

This hit me like a bolt of lightning. For the first time I realized the obvious: This miraculous chip was a man-made miracle. All the marvels of the computer age, all the "electronic brains" and "artificial intelligence," were simply products of the most powerful intelligence of all—the human brain. Beyond that, I was quite taken with the concept that an idea could have a name—"the monolithic idea"—and that two living Americans had conceived an idea that would change the world.

I resolved to meet the men responsible for the invention and find out how they had done it. This book is the result. The first edition came out just as a new consumer product, the personal computer, was bringing the information age into millions of homes and offices for the first time. Back then, it seemed to me just a strange quirk that my fellow Americans were not yet aware of the two countrymen who had launched this new era. It seemed perfectly obvious—to me, at least—that Noyce and Kilby would soon be household names along with Edison, Bell, and Ford. That didn't happen. This second edition of the book, with the story considerably expanded and carried forward to a new century, is another effort to suggest that these two men are modern heroes who should be admired and emulated around the world.

Many people have helped me with this book, and my gratitude is enormous. My deepest thanks go to Jack Kilby and Robert Noyce, who were generous with their time and unfailingly patient with my dumb questions. Their colleagues Willis Adcock, Melvin Sharp, Mac Mims, Jerry Merryman, Norman Neureiter, Dick Perdue, Gordon Moore, Jean Hoerni, and Roger Borovoy went out of their way to help me, as did the librarians at both Texas Instruments and Intel Corp. Howard Warshaw of Atari Corporation took me on a guided tour through the inside of a chip. Homer Reid of Bell Labs steered me expertly through the intricacies of quantum theory and digital circuit design. Georgine Neureiter, Tom and Janet Cameron, Jane Kilby, and Ann Kilby have all offered invaluable aid, for no reason that I can think of other than sheer kindness.

I have relied on two great national collections of knowledge. The Science Reading Room at the Library of Congress is an American treasure; the knowledgeable staff there were the mainstays of my research. The Science Collection of the British Library generously made its copious materials available. The libraries of Princeton, Georgetown, and Stanford Universities also permitted me to use their resources. I owe thanks to the public libraries of Denver, Colorado, Washington, D.C., and the Borough of Westminster in London—and particularly to the P. S. Miller branch of the Douglas County Library in the foothills town of Castle Rock, Colorado. The Cleveland Institute of Electronics allowed me the use of educational materials. Professor Elizabeth Tuttle of the Physics and Engineering Department at the University of Denver and Jeff Singh of the Mathematics Department there provided valuable help.

Maralee Schwartz, one of the world's great researchers and editors, demonstrated time and again that no fact was too obscure for her to find. I once asked her to track down the speed at which an eyelid blinks; she was back in the wink of an eye with a detailed report. In London, I was lucky to benefit from the awesome reportorial talent of Adi Bloom. Many U.S. government officials, including John McPhee of the Commerce Department, Isaac Fleischmann of the Patent Office, and Dr. Uta C. Merzbach of the Smithsonian Institution, were helpful along the way.

Several of my colleagues in Washington assisted with this book. The charming bon vivant Ben Bradlee and the cheery iconoclast Bill Greider both recognized early on that this was an important piece of contemporary history and that I ought to pursue it. Nick Lemann, Joel Garreau, and Margaret Shapiro provided thoughtful and useful advice. I am indebted to Mary McGrory, Patrick Gross, Haynes Johnson, and Admiral H. G. Rickover for recognizing from the beginning that this effort would turn into something worthwhile. I'm grateful to Bob Shrum and Dr. William Leahy for solid advice at the end. I've had the good fortune to work with two wonderful editors on one book: Alice Mayhew and Ann Godoff. They figured out where the book should be heading and helped get me there. Henry Ferris and Kate Niedzwiecki offered smart and patient editorial guidance. Christopher McLehose of Collins, Ltd., provided wry advice. My agents, first Rhoda Weyr and now Gail Ross, were there for me to lean on whenever I needed them.

In its first realization, this book was written on a Heathkit H-89 computer based on an early microprocessor known as the Z-80. In those ancient times before Windows, the H-89 used a now-defunct operating system called CP/M and a now-defunct word processing program called Peachtext, both of which worked pretty well. I wrote this new edition on a Hewlett-Packard Pavilion PC built around Intel's Pentium III processor.

My education in electronics and my first year of research on this book were supported by the Alicia Patterson Foundation. I owe particular thanks to Alice Arlen, Joe Albright, Helen Coulson, and Cathy Trost at the foundation. The Washington Post Company and its estimable chief, Don Graham, gave me the time and space to write the book not once but twice during my *Post* career.

Last but foremost, Margaret Mary McMahon, McMahon Thomas Homer Reid, O'Gorman Catherine Penelope Reid, and Erin Andromache Wilhelmina Reid put up with me and the manuscript in cheery fashion, a task far more formidable than writing any book.

Washington, D.C.   London
1981   2001

A NOTE ABOUT SOURCES

Readers who plan to dig deeper into any of the events or characters of this story have a treat in store. There are some wonderful books and Web sites about the physics and the engineering behind the digital revolution. However, if you stop by a bookstore, library, or search engine and just start browsing, you'll find there's much more in print or on screen than anyone could read. Accordingly, I have compiled the following road map to steer you toward some of the better sources I ran across while writing this one.

The Royal Swedish Academy of Sciences loosed a flood of information about "Physics and Information Technology" into the bit stream when it announced Jack Kilby's Nobel Prize. This can be found at www.nobel.se/physics/laureates/2000/illpres/kilby/html. A related source is an excellent Web site called The Nobel Prize Internet Archive, at www.almaz.com/nobel/nobel.html.

The best nontechnical book I found on the general history of semiconductor electronics (although it is somewhat skimpy on the invention and development of the chip) is *Revolution in Miniature,* by Ernest Braun and Stuart MacDonald (Cambridge, U.K.: Cambridge University Press, 1983). S. Handel, *The Electronic Revolution* (Baltimore: Penguin, 1979), is an easy and in-

teresting discussion of electronics history from Benjamin Franklin
to the invention of the transistor. The basic text on the tyranny of
numbers was written by the man who coined that term, Jack A.
Morton, *Organizing for Innovation* (New York: McGraw-Hill,
1971). There are also two big, expensive coffee-table books on the
history of electronics: The Editors of Electronics, *An Age of Innovation: The World of Electronics, 1930–2000* (New York: McGraw-
Hill, 1981), which looks at the story through American eyes; and
Elizabeth Antébi, *The Electronic Epoch* (New York: Van Nostrand
Reinhold, 1982), which provides a European view of the same period.

Somewhat more technical, but still accessible to the lay reader,
is Scientific American, *Microelectronics* (San Francisco: W. H.
Freeman, 1977), which includes an overview of this technical revolution written by Robert Noyce. During the bicentennial year,
several of the technical journals of the IEEE (the acronym is pro-
nounced "I triple E" and stands for the Institute of Electrical and
Electronic Engineers) gave a great deal of attention to technical
history. The July 1976 edition of *IEEE Transactions on Electron
Devices*, available at most large public or research libraries, contains first-person accounts of important inventions by many
preeminent engineers, including William Shockley (on the transistor) and Jack Kilby (on the chip).

Several reference works focus on the matters covered in this
book. I grew to rely on the Van Nostrand *Encyclopedia of Computer Science* (1976) and the fifteen-volume McGraw-Hill *Encyclopedia of Science and Technology* (1982). G.W.A. Dummer,
Electronic Inventions and Discoveries (New York: Pergamon,
1983), is a listing of important developments. For information on
individual scientists and engineers, Isaac Asimov, *Asimov's Biographical Encyclopedia of Science and Technology* (Garden City,
N.Y.: Doubleday, 1982), is a delightful quick resource. The mul-
tivolume *Dictionary of Scientific Biography* (New York: Scribner,

1970), edited by Professor Charles C. Gillispie, is a classic piece of scholarship that provides clear and comprehensive biographies of hundreds of scientists and engineers.

Two fairly recent biographies of Thomas A. Edison, Matthew Josephson, *Edison* (New York: McGraw-Hill, 1959), and Robert Conot, *A Streak of Luck* (New York: Seaview, 1979), are well done and rich with detail. The best source on Francis Upton's relationship with the Wizard of Menlo Park is Upton's own history, *Edison's Electric Light Bulb* (1881). For a pleasant history of physics in England in J. J. Thomson's time, there is J. G. Crowther, *The Cavendish Laboratory 1874–1974* (New York: Science History Publications, 1974). J.J.'s son, the Nobel laureate Sir George P. Thomson, wrote a fascinating memoir of his father, *J. J. Thomson and the Cavendish Laboratory in His Day* (Garden City, N.Y.: Doubleday, 1965). The most delightful source of information on J. J. Thomson, though, is his modest, intriguing 1936 autobiography, *Recollections and Reflections*, which has been reprinted by Arno Press (New York, 1975).

John A. Fleming wrote a memoir of his own, *Fifty Years of Electricity* (New York: Wireless Press, 1921), and there is also a short remembrance by his devoted lab assistant, J. T. McGregor-Morris, *The Inventor of the Valve* (London: The Television Society, 1954). Georgette Carneal's *A Conqueror of Space* (New York: H. Liveright, 1930) was an authorized biography of Lee De Forest. Twenty years later, though, the inventor produced the autobiography with the immodest title mentioned in Chapter 2: Lee De Forest, *Father of Radio* (Chicago: Wilcox & Follett, 1950). C. P. Snow's *The Physicists* (Boston: Little, Brown, 1981) contains a warm portrait of Niels Bohr.

William Shockley's definitive *Electrons and Holes in Semiconductors, with Applications to Transistor Electronics* (New York: Van Nostrand, 1950) is the basic text on semiconductor physics and on Shockley's way of thinking; for the first hundred pages or so, it is accessible to any diligent reader. The invention of the transistor has been related by the inventors in a series of journal articles and in the lectures they delivered upon receiving the Nobel Prize; many can be found on the Nobel Foundation's Web site, www.nobel.se. A National Geographic book, *Those Inventive Americans* (Washington, D.C.: National Geographic Society, 1971), provides a popularized look at the transistor and its fathers; a somewhat more technical presentation is offered in George L. Trigg, *Landmark Experiments in 20th-Century Physics* (New York: Crane, Russak, 1975).

Except for this book, there is very little between covers about Jack Kilby and Robert Noyce. Tom Wolfe wrote a terrific article about Noyce, "The Tinkerings of Robert Noyce," for *Esquire* magazine in December 1983, and it appears in his collection *Hooking Up* (New York: Farrar, Straus and Giroux, 2000). The German Hans Queisser takes a curiously Eurocentric view of the story in *The Conquest of the Microchip* (Cambridge, Mass.: Harvard University Press, 1988). Most of the other recent books dealing with the history of the chip appear to have borrowed shamelessly from this one; some mention this book as a source, some do not. Texas Instruments, with justified pride, maintains a good Jack Kilby archive on its Web site at www.ti.com/corp/docs/kilbyctr/jackstclair.shtml.

A reader who would like to delve deeper into patent law can obtain a number of interesting and instructive pamphlets from the Patent Office Web site, www.uspto.gov. For more detailed historical and legal information, the best one-volume source I have found is Peter D. Rosenberg, *Patent Law Fundamentals* (New York: Clark Boardman, 1975).

The binary system and other mathematical principles underlying digital computer operations are discussed in the four-volume *The World of Mathematics* by James R. Newman (New York: Simon & Schuster, 1956); volume 3 of this excellent set also includes a useful section on Boolean logic. The reader who is interested in math, though, will derive the most pleasure from the witty, insightful, and generally marvelous books of Eric Bell: two that pertain directly to material covered in this book are *Mathematics, Queen and Servant of Science* (New York: McGraw-Hill, 1951) and *Men of Mathematics* (New York: Simon & Schuster, 1937); the latter includes a fine portrait of George Boole. There is also interesting Booleana in Mary Everest Boole, *A Boolean Anthology* (Association of Teachers of Mathematics, 1972). Dover Press deserves our gratitude for keeping in print a paperback version of George Boole's masterpiece, *The Laws of Thought* (New York: Dover Publications, 1951). There is as yet no biography of Claude Shannon, but a reader might be interested in the book that launched the burgeoning field of information theory—that is, Claude E. Shannon, *The Mathematical Theory of Communication* (Champaign: University of Illinois Press, 1949).

Computer history is just now emerging as an academic discipline of its own, and there will no doubt be some fine books written on the work of von Neumann, Turing, and other computer pioneers. There is a good general history in Joseph C. Giarratano, *Foundations of Computer Technology* (Indianapolis: Howard W. Sams, 1982). An important contribution to this literature is Herman Goldstine's *The Computer from Pascal to von Neumann* (Princeton, N.J.: Princeton University Press, 1972), which is strangely organized but has the immediacy that could be conveyed only by one who was present at the creation of the modem electronic computer. Andrew Hodges, *Alan Turing: The Enigma* (New York: Simon & Schuster, 1983), and Steve J. Heims, *John von Neumann and Norbert Wiener* (Cambridge, Mass.: MIT Press, 1980), are the first complete biographies. Von Neumann's seminal paper "Preliminary Discussion of the Logical Design of an Electronic Computing Instrument" is reprinted in John Diebold, ed., *The World of the Computer* (New York: Random House, 1973).

There are far more books than any one person could read on the inner workings of integrated circuits, microprocessors, calculators, and computers. Professor Jacob Millman's standard text, the one that got me started, has been updated by a colleague: Jacob Millman and Arvin Grabel, *Microelectronics* (New York: McGraw-Hill, 1987). It is excellent but can be heavy going for the non-engineer. At the other end of the scale is Larry Gonick, *The Cartoon Guide to Computer Science* (New York: Barnes & Noble, 1983), which manages to be hilarious and quite informative at the same time. The IEEE put out a good primer on logic circuits, John Gregg, *Understanding Boolean Algebra, Digital Circuits and the Logic of Sets* (Los Alamitos, Calif.: IEEE Press, 1998).

The DIM-I calculator in this book is based to some degree on the "Simple as Possible" computer designed by Albert Paul Malvino in *Digital Computer Electronics: An Introduction to Microprocessors* (New York: McGraw-Hill, 1983). Other books describing how a computer gets the answer include Gene McWhorter, *Understanding Digital Electronics* (Dallas: Texas Instruments Learning Center, 1978); Rodnay Zaks, *From Chips to Systems: An Introduction to Microprocessors* (Alameda, Calif.: Sybex, 1981); and the three-volume series by Adam Osborne, *An Introduction to Microcomputers* (Berkeley, Calif.: Osborne/McGraw-Hill, 1982).

The flood of books on Japanese management that has poured forth upon this country in recent years includes several that discuss Japanese competition in semiconductor electronics. An excellent case study of Japanese success is Paul duGay et al., *The Story of the Sony Walkman* (Thousand Oaks, Calif.: Sage, 1997). The key source on that intriguing figure W. Edwards Deming is Dr. Deming's own text, *Quality, Productivity, and Competitive Position* (Cambridge, Mass.: MIT Press, 1982).

For those who want to know where the digital revolution is heading, Richard Turton's *The Quantum Dot* (New York: Oxford University Press, 1995) offers a fairly technical description of how much further the integrated circuit can be integrated, and what technology is available when semiconductor chips are finally saturated. The economics of the digital future are pondered in interesting fashion by George Gilder in *Microcosm: The Quantum Revolution in Economics and Technology* (New York: Simon & Schuster, 1989).

NOTES

(Fuller publishing information appears in A Note about Sources.)

Chapter 1: The Monolithic Idea

3 honored in the textbooks: Cf. Jacob Millman, *Microelectronics*, p. xxii.

3 willing to overlook: "The Nobel Prize in Physics, 2000," Royal Swedish Academy of Sciences, Oct. 10, 2000.

3 he only hoped . . . Jack had been delighted . . . some misgivings: Interview with Jack Kilby.

4 reduced titanate capacitor: U.S. Patent No. 2,841,508 (1958).

4 steatite-packaged transistor: Kilby, "Transistor Amplifier Packaged in Steatite," *Electronics*, October 1956.

4 It was that infatuation: Interview with Kilby.

4 the technical journals called it: Cf. Robert Noyce, "Interconnections," *Proceedings of the IEEE*, December 1964, p. 1648 ff.; Jack A. Morton, "The Microelectronics Dilemma," *International Science & Technology*, July 1966, p. 38.

5 the "neuristor": G.W.A. Dummer, "Progress with Extremely Small Elec. Circuits," *New Scientist*, Feb. 7, 1963, pp. 283–84.

5 a *Life* magazine reporter: *Life*, Nov. 5, 1956.

5 was said to be working on: Interview with Kilby.

6 "He was enthused, but . . .": Kilby, "Invention of the Integrated Circuit," *IEEE Transactions on Electron Devices*, July 1976, p. 650.

6 "I was very . . .": Interview with Willis Adcock.

8 "If we built . . .": Interview with Robert Noyce.

9 "... it would be desirable ...": Noyce's lab notebook for January 1959 (copy in author's file); also "The Genesis of the Integrated Circuit," *IEEE Spectrum*, August 1976, p. 47.

9 "There was a tremendous ...": Interview with Noyce.

10 about 10,000 times in a second: Cf. *Scientific American*, January 1982, p. 124.

10 known in England: Ernest Braun and Stuart MacDonald, *Revolution in Miniature*, p. 16.

11 debugging: Joseph C. Giarratano, *Foundations of Computer Technology*.

12 The first transistor radio: Cf. *Circuit News*, Apr. 15, 1979.

12 "Interpretation of ...": *Engineering Index*, 1953.

12 "Le Transistron ...": Ibid., 1955.

12 "Circuito ...": Ibid.

12 "Tensoranalysis ...": Ibid.

12 "Perekhodnaya ...": Ibid., 1958.

13 "Success Story ...": Ibid., 1957.

13 "Transistors Key ...": Ibid., 1953.

13 "Méthodes ...": Ibid., 1962.

13 "Fabulous ...": Ibid., 1953.

13 "Switching ...": Ibid., 1962.

13 "Electronic ...": Ibid., 1961.

13 "Comment ...": Ibid., 1959.

13 "Design ...": Ibid., 1960.

15 had 350,000 electronic: Horace D. Gilbert, ed., *Miniaturization* (New York: Reinhold, 1961), p. 3.

15 the Control Data CD 1604: Interview with Kilby; *Electronics*, April 1980, p. 81.

16 "For some time now ...": Jack A. Morton, in *Proceedings of the IRE*, June 1958, p. 955 ff.

16 "Each element must ...": Morton, in *International Science & Technology*, July 1966, p. 38.

17 Royal Radar Establishment: Braun and MacDonald, p. 108.

17 "tend to exacerbate ...": Patrick Haggerty, in *Proceedings of the IEEE*, December 1964, p. 1648.

17 under the general title: *Electronics News*, Jan. 25, 1982, p. 16.

17 Operation Tinkertoy: Kilby, in *Transactions on Electron Devices*, July 1976, p. 648.

17 "In civilian equipment ...": *Electronics*, Oct. 1, 1957, p. 178.

18 designers tried redundancy: Cf. Kilby, in *Electronics*, Aug. 7, 1959, p. 111.

19 "After you become ...": Interview with Robert Noyce.

21 "It was a situation ...": Ibid.

21 "The things that ...": Interview with Kilby.

21 "A large segment . . .": Noyce, in *Scientific American*, September 1977, p. 64.

21 "There was just . . .": Interview with Kilby.

21 "It was clear . . .": *Scientific American*, September 1977, p. 64.

22 Richard Feynman put it: Feynman, *The Pleasure of Finding Things Out*, (Cambridge, Mass.: Perseus, 1999), chap. 1.

23 "Second Industrial Revolution": National Academy of Sciences, *Microstructure Science, Engineering & Technology*, 1979, p. 1.

23 "the most remarkable . . .": In Tom Forester, ed., *The Microelectronics Revolution*, (Cambridge, Mass.: MIT Press, 1981), p. 1.

23 "Integrated circuits are . . .": Interview with Noyce.

Chapter 2: The Will to Think

24 observed for the first time in March 1883: Robert Conot, *A Streak of Luck*, p. 337; Isaac Asimov, *Understanding Physics*, vol. 3 (New York: Walker, 1966), p. 42.

25 he saw no future in it: Conot, p. 337 ff.; Asimov, vol. 3, p. 42.

25 "the zero hour . . .": Sir George P. Thomson, *J. J. Thomson*, p. 20.

26 "Well, I'm not a scientist . . .": Matthew Josephson, *Edison*, p. 283.

26 Dot and Dash: Conot, p. 122.

26 *New York Daily Graphic*: Ibid., p. 113.

26 When Edison died: Josephson, pp. 484–85.

27 "The hair, beginning . . .": Ibid., p. 132.

27 *Perseverantia omnia vincit* . . . ". . . 99 percent perspiration": Conot.

27 "at least we know 8,000 . . .": Ibid.

27 "filled up with Latin . . ."; "ignoramer": Josephson, pp. 440, 442.

27 named him "Culture": Conot, p. 133.

29 quickly dubbed the Edison Effect: Josephson, p. 278. Cf. Asimov, vol. 3, p. 42.

29 "I have never . . . the Savanic world": Josephson, p. 278; Conot, p. 337.

31 we still talk today of "current": S. Handel, *The Electronic Revolution*, pp. 20–21.

31 "that in a few years all the great . . .": J. G. Crowther, *The Cavendish Laboratory*, p. 38.

32 Trinity Prize; 1881 paper; succeeded in 1884: Crowther; G. Thomson, pp. 20–24; J. J. Thomson, *Recollections & Reflections*.

32 "Things have come to a pretty pass . . .": Crowther, p. 109.

32 "J.J. spent a good part of most days . . .": G. Thomson, p. 114.

32 In his memoirs: J. J. Thomson, p. 435.

33 "In the dusty lab'ratory, / 'Mid the coils and wax and twine . . .": G. Thomson, p. 98.

35 "I have lately made . . ."; "corpuscles": Bragg and Porter, eds., *Royal Institute Library of Science*, pp. 43, 48.

37 Fleming was determined . . . "There was a young fellow . . ." . . . "scientific hooliganism": J. T. McGregor-Morris, *The Inventor of the Valve*, pp. 62, 66.

39 came to be known as crystal sets: Cf. Asimov, vol. 3, p. 95.

39 I was pondering . . .: John A. Fleming, *Popular Radio*, Mar. 19, p. 175.

40 "So nimble are . . .": Fleming, *Fifty Years of Electricity*, p. 335.

41 "absurd and deliberately . . .": Gerald F. J. Tyne, *Saga of the Vacuum Tube* (Indianapolis: Sams, 1977).

41 "a more revolutionary step . . .": Georgette Carneal, *A Conqueror of Space*, p. 187.

41 "few inventions have had so many fathers": Isaac Asimov, *Asimov's Biographical Encyclopedia of Science and Technology*.

41 "Commonly known as . . .": *Electronics News*, Jan. 25, 1982, sec. 2, p. 10.

42 "grid"; "Device for Amplifying . . .": Asimov, *Understanding Physics*, vol. 3, p. 42; Carneal; Tyne.

42 iconoscope: Cf. Asimov, *Understanding Physics*, vol. 3, p. 50.

43 "radar" . . . Crystal rectifiers: Cf. Handel, pp. 80–81.

45 "the incarnation . . ."; "He was not, as Einstein was . . .": C. P. Snow, *The Physicists*, p. 51 ff.

46 *Contraria sunt complementa*: John Daintith, Sarah Mitchell, and Elizabeth Tootill, eds., *A Biographical Encyclopedia of Scientists*, (New York: Facts on File, 1981), p. 87.

47 "In this picture we see a striking . . .": *Nobel Lectures—Physics, 1922–41*, p. 8.

49 "the vacant parking place . . .": William Shockley, *Electrons and Holes in Semiconductors*, p. 8.

50 N-type; P-type: Cf. Jacob Millman, *Microelectronics*, pp. 12–13.

51 "It has today occurred to me . . .": Shockley, "The Path to the Conception of the Junction Transistor," *IEEE Transactions on Electron Devices*, July 1976, p. 603.

51 "Try simplest cases"; "the will to think"; "In these four words . . .": Ibid.; pp. 600, 599–600.

51 "the wave function A (ϕ) for the hole-wave packet . . .": Shockley, *Electrons and Holes in Semiconductors*, p. 444.

52 taught a freshman seminar: "Brave New William Shockley," *Esquire*, January 1973, p. 130 ff.

52 "THINKING about THINKING . . .": *IEEE Student Journal*, September 1968, pp. 11–16.

52 "dysgenics"; "retrogressive evolution . . .": *Esquire*, January 1973. Cf. *The New York Times*, May 3, 1970, p. 58, and Jan. 17, 1970, p. 30 (letter to editor from Shockley).

53 "decline of our nation's human quality": *The New York Times*, Jan. 17, 1970, p. 30.

53 "My research leads me inescapably . . .": Ibid., Dec. 5, 1973, p. 38.

53 a social policy to deal with: Ibid., Dec. 13, 1973, p. 95.

53 "Off Pig Shockley"; photograph; microphone went on the blink: *Esquire*, January 1973, p. 130 ff.

54 pursued the Republican nomination: *The New York Times*, Feb. 12, 1982, p. 16.

54 "charming": Interview with Robert Noyce.

54 "the vacuum tube and thermionics . . ."; "Don't worry, Walter . . .": "Walter Houser Brattain," Bell Laboratories Record, December 1972, p. 339 ff.

55 "one indeed needs to be . . .": *Nobel Lectures—Physics, 1942–62*, (New York: Elsevier, 1964), p. 377 (Brattain).

55 "I feel strongly, however . . ."; "The thing I deplore . . .": *Bell Laboratories Record*, December 1972, p. 339.

55 "if I wiggled it just right": George L. Trigg, *Landmark Experiments in 20th-Century Physics*, p. 149.

57 By 1946 . . . the operation of the semiconductor diode: *A History of Technology in the 20th Century*, pt. 2, p. 1117.

58 "Hearing speech amplified . . .": Shockley, in *IEEE Transactions*, July 1976, p. 611.

59 "My elation with the group's success . . .": Ibid., p. 612.

Chapter 3: A Nonobvious Solution

63 ". . . a couple of slides": Interview with Dr. Tord Claesson, Swedish Academy.

64 Quotations from Kilby: Interviews with Jack Kilby.

65 "The definition of the problem . . ."; "Although invention is . . .": Kilby, "The Inventor's View," *Chemtech*, March 1979, p. 66.

66 Quotations from Kilby: Interviews with Kilby; ibid.

73 Geophysical Research Corporation: "The Men Who Made T.I.," *Fortune*, November 1961.

74 "Our company now has two . . .": Interview with Willis Adcock. Cf. *Fortune*, November 1961.

76 "I felt it likely . . .": Kilby, "Invention of the Integrated Circuit," *IEEE Transactions on Electron Devices*, July 1976, p. 650.

76 "If Texas Instruments . . .": Interview with Kilby.

77 "The following circuit elements . . .": Kilby's lab notebook, July 24, 1958 (copy in author's file).

78 "Nobody would have made . . .": Interview with Kilby.

78 Quotations from Kilby: Ibid.
78 "it was pretty damn cumbersome": Interview with Adcock.

Chapter 4: Leap of Insight

81 Quotations from Noyce: Interview with Robert Noyce.
83 "They're high achievers . . .": *Harvard Business Review*, May/June 1980, p. 122.
83 Quotations from Noyce: Interview with Noyce.
87 low-energy gas discharges: *Journal of Applied Physics*, vol. 27, August 1956, p. 843.
87 "dc transistor current amplification factor": *Proceedings of the IRE*, September 1957, pp. 1228–43.
87 "base widening punch-through": Interview with Noyce.
88 Semiconductor diffusion . . . 1000 degrees centigrade or so: Jacob Millman, *Microelectronics*, p. 100.
89 described Noyce as an introvert: *The Economist*, Dec. 27, 1980, p. 63.
90 "There was a group . . .": Interview with Noyce.
90 "the traitorous eight": Dirk Hanson, *The New Alchemists* (Boston: Little, Brown, 1982), p. 92; also, interview with Gordon Moore.
90 "the realization . . .": Interview with Noyce.
91 the return on his initial $500: Gene Bylinsky, *The Innovation Millionaires* (New York: Scribner, 1976), p. 61 ff.
91 "Wall Street has spoken . . .": *Financial World*, Mar. 15, 1981, p. 30 ff.
91 "Here we were . . .": Interview with Noyce.
92 Quotations from Noyce, Ralls: Ibid.
94 "all the bits and pieces . . .": "The Genesis of the Integrated Circuit," *IEEE Spectrum*, August 1976, p. 50.
94 "In many applications . . .": Noyce's lab notebook for January 23, 1959 (copy in author's file).
95 "a unitary circuit structure . . .": U.S. Patent No. 2,981,877, p. 1.

Chapter 5: Kilby v. Noyce

(Note: "Docket" refers to the file of documents submitted for Interference Proceeding No. 92,841, Kilby v. Noyce, at the U.S. Patent Office, Arlington, Virginia.)
96 The terrifying rumor: Testimony of Martin Fleit, July 1964—transcript in docket. Also, interview with Melvin Sharp, patent counsel, TI.
97 Mosher promised: Interview with Sharp.

97 "the right to exclude . . .": U.S. Code—cited in *General Information Concerning Patents* (Washington, D.C.: U.S. Department of Commerce, 1982), p. 2.

97 "a written description . . .": 35 U.S.C. 112.

98 Coca-Cola history: *The Coca-Cola Co.—An Illustrated Profile* (Atlanta: Coca-Cola Co., 1974); Charles Howard Candler, *Asa Griggs Candler* (Atlanta: Emory University Press, 1950), p. 119.

98 trade secret protection of Coca-Cola formula: Peter D. Rosenberg, *Patent Law Fundamentals*, pp. 15–16.

99 shredded wheat case: *Kellogg Co. v. National Biscuit Co.*, 305 U.S.C. 111 (1938).

100 "The sheets may be provided . . .": *General Information Concerning Patents*, p. 17.

100 "new"; "useful": 35 U.S.C. 101; no patent on nuclear weapons: 42 U.S.C. 2181.

102 That first telephone patent: Rosenberg, p. 12.

102 "Radically departing . . ." and other quotations from patent application: U.S. Patent No. 3,138,743.

103 ". . . shall set forth the best mode . . .": 35 U.S.C. 112.

104 *integration* and *interconnection*: Interview with Jack Kilby; Kilby, "Invention of the Integrated Circuit," *IEEE Transactions on Electron Devices*, July 1976.

105 "flying wire picture": Interview with Roger Borovoy.

106 "Instead of using . . .": U.S. Patent No. 3,138,743.

108 "We were still a brand-new . . .": Interview with Robert Noyce.

109 Quotations from patent application: U.S. Patent No. 2,981,877.

110 interference proceeding; Board of Patent Interferences: 35 U.S.C. 13(a).

110 received a copy of . . . Form POL-102: In docket.

111 "Motion to Dissolve . . ." and other documents: Ibid.

112 "Here I was . . .": Interview with Borovoy.

113 "conducting material . . .": U.S. Patent No. 3,138,743.

113 experts' testimony: Testimony of Professor G. Pearson (Stanford), Oct. 9, 1964; testimony of Professor R. Maurer (Illinois), Nov. 24, 1964 (transcripts in docket).

114 "Request for . . ." and other motions: In docket.

114 Borovoy's brief included; "Note that this . . .": Copy of brief in docket.

114 "we are not particularly . . .": Opinion of the Patent Office Board of Patent Interferences, No. 92,841, Feb. 24, 1967 (in docket).

115 CCPA decision: *Kilby v. Noyce*, 416 F2d. 1391 (CCPA, 1969).

115 "Denied": *Kilby v. Noyce*, cert. denied Oct. 12, 1970 (in docket).

115 in the summer of 1966: Interviews with Sharp and Borovoy; *Electronics News*, May 9, 1966, p. 1.

117 "Patent Appeals Court Finds . . .": Cf. *Electronics News*, Nov. 24, 1969, p. 63.

Chapter 6: The Real Miracle

118 March 24, 1959; "solid circuits": Interviews with Jack Kilby and Willis Adcock. Cf. *Electronics*, Mar. 13, 1959.
118 a lavish prediction . . . from TI's president: Editors of Electronics, *An Age of Innovation*, p. 84.
118 "It wasn't a sensation": Interview with Kilby.
118 In its special issue: *Electronics*, Mar. 13, 1959.
119 "match-head size solid-state circuit": Ibid., Apr. 3, 1959, p. 11.
119 "There was a lot of flak . . .": Interview with Kilby.
119 The critics identified three; "These objections were . . .": Kilby, "Invention of the Integrated Circuit," *IEEE Transactions on Electron Devices*, July 1976, p. 652.
119 one common line of analysis: Interview with Robert Noyce. Cf. Noyce, "Microelectronics," *Scientific American*, September 1977, p. 67.
119 the giants of the industry . . . kept themselves clear: Ibid.; also, Noyce, *Scientific American*, September 1977, p. 68.
120 "The synergy between a new component . . .": Noyce, *Scientific American*, September 1977, p. 63.
121 traditional economies were reversed: Ibid.
122 *On Beyond Zebra*: Dr. Seuss, *On Beyond Zebra* (New York: Random House, 1955).
123 positional notation; invention of a symbol: Isaac Asimov, *Asimov on Numbers* (Garden City, N.Y.: Doubleday, 1977), pp. 21–28.
127 ones-complement subtraction: Albert Paul Malvino, *Digital Computer Electronics*, pp. 81–84.
129 "Accused Had Powerful Brain": Andrew Hodges, *Alan Turing: The Enigma*, p. 474.
129 "One day ladies will take . . ."; "in about fifty years' time.": Ibid., pp. 418, 417.
130 The filmmaker Stanley Kubrick: Ibid., p. 533.
130 Turing wrote, it was natural enough: Ibid., p. 320.
130 "We feel strongly in favor . . .": John von Neumann et al., "Preliminary Discussion of the Logical Design of an Electronic Computing Instrument," in John Diebold, ed., *The World of the Computer*, p. 54.
130 One was that the human race: Phillips, "Binary Calculation" (1936), in Brian Randell, ed., *The Origins of Digital Computers* (New York: Springer-Verlag, 1975), p. 293.

131 "The one disadvantage of the binary system . . .": Diebold, pp. 55–56.

131 decoder: Eugene McWhorter, *Understanding Digital Electronics*, pp. 8–12.

132 "flash of psychological insight": Mary E. Boole, *A Boolean Anthology*, pp. 61, 62; William Kneale, "Boole and the Revival of Logic," *Mind*, vol. 57 (April 1948), p. 152.

133 "that his children not be allowed . . .": Kneale, quoted in James R. Newman, *The World of Mathematics*, vol. 3, p. 1854.

133 "The design of the following treatise . . .": George Boole, *The Laws of Thought*, p. 1.

134 "Sonnet to the Number Three": Kneale, vol. 57.

134 "No mere mathematician can understand . . .": M. Boole, p. 68.

135 $x(1 - y)(1 - z) + y(1 - x)(1 - z)$ . . .: G. Boole, *The Laws of Thought*.

135 $x = x^2$: Ibid., p. 64.

136 " 'You are sad,' the Knight said . . .": Lewis Carroll, *Through the Looking Glass*.

136 In *Principia Mathematica*: Newman, vol. 3, pp. 1894–95.

136 "The abstract doctrines . . .": G. Boole.

136 Vannevar Bush, had designed: Herman Goldstine, *The Computer from Pascal to von Neumann*, p. 90.

138 "Programming a Computer for Playing Chess": *Philosophical Magazine*, vol. 41 (1950), p. 256.

138 "Squdgy fez, blank jimp crwth vox!": *The New York Times*, Aug. 16, 1972, p. 36.

139 "It is possible to perform . . .": Claude Shannon, "A Symbolic Analysis of Relay and Switching Circuits," *AIEE Transactions*, vol. 57 (1938), p. 722.

140 "In fact, any operation . . .": Ibid.

142 gates: Cf. Jacob Millman, *Microelectronics*, p. 123.

Chapter 7: Blasting Off

144 a new line of six different . . . "Micrologic elements": *Electronics*, Mar. 31, 1961, p. 91.

144 a similar series of "solid circuits": Interview with Jack Kilby; Editors of Electronics, *An Age of Innovation*, p. 83.

144 priced at $120: *Electronics*, Mar. 31, 1961, p. 76.

144 "There was the natural . . .": Interview with Robert Noyce.

145 learning curve: Ernest Braun and Stuart MacDonald, *Revolution in Miniature*, p. 82.

146 "The space program badly needed . . .": Interview with Jack Kilby.

146 U.S. electronics companies complained: Cf. "Competitive Factors Influencing World Trade in Semiconductors," Hearings before U.S. House Subcommittee on Trade, No. 96–62, Nov. 30, 1979.

147 A study published in 1977: Linville and Hogan, in *Science*, Mar. 18, 1977, p. 1109.

147 federal government remained the largest buyer: *Forbes*, Feb. 15, 1971, p. 21; Braun and MacDonald, pp. 141–42.

147 "The general rule of thumb . . .": Interview with Noyce.

147 specifications called for every single component: Interview with Willis Adcock; Braun and MacDonald, p. 114.

147 Testing, retesting, and re-retesting: Jack A. Morton, *Organizing for Innovation*, p. 106; *Business Week*, Apr. 14, 1962, p. 168.

147 services went off in three different directions: Kilby, "Invention of the Integrated Circuit," *IEEE Transactions on Electron Devices*, July 1976, p. 649.

148 "molecular electronics": Cf. Air Force Systems Command, *Integrated Circuits Come of Age* (United States Air Force, 1966), p. 4.

148 "The idea of it was . . .": Interview with Noyce.

149 "TI has always followed a strategy . . .": Interview with Kilby.

149 "The development of integrated circuits . . .": *Integrated Circuits Come of Age*, p. 1.

149 "almost an insult"; "The missile program and the . . .": Interview with Noyce.

150 The designers of Minuteman II . . . 4,000 chips per month: *Integrated Circuits Come of Age*, p. 16; *An Age of Innovation*, p. 325.

150 About 500,000 . . . were sold . . . and quadrupled again: Kraus, *An Economic Study of the U.S. Semiconductor Industry*, pp. 80, 209.

151 "From a marketing standpoint . . .": Interview with Kilby.

151 The first chip sold for the commercial market: *Integrated Circuits Come of Age*, p. 20.

152 In 1963 the price of an average chip: "A Report on the U.S. Semiconductor Industry," U.S. Commerce Department 99, p. 50; Electronic Industries Association, *Market Data Book 1975*, p. 84.

152 "A single speck of dust . . .": Interview with Noyce.

153 "I did it sort of tongue-in-cheek . . .": Interview with Gordon Moore.

154 256M chip: "Physics and the Information Revolution," *Physics Today*, January 2000.

154 Pentium IV: Ibid.

156 "We would cruise comfortably . . .": Moore, "The Cost Structure of the Semiconductor Industry," *IEEE Transactions Consumer Electronics*, February 1977.

156 "Progress has been astonishing . . .": Noyce, "Microelectronics," *Scientific American*, September 1977, p. 65.

157 IBM brought out . . . System 360 . . . competitors had to: *An Age of Innovation*, p. 339 ff.

159 A monograph that appeared: David A. Hodges, "Large-Capacity Semiconductor Memory," *IEEE Proceedings*, July 1968, p. 1148.

161 Intel rang up a grand total: *Fortune*, November 1973, p. 142.

161 linear, or analog, integrated circuit: Jacob Millman, *Microelectronics*, p. 333.

162 "We reached a point . . .": Interview with Moore.

163 To maintain its explosive rate of growth: Cf. Moore, "The Cost Structure of the Semiconductor Industry."

Chapter 8: The Implosion

164 "Pervasiveness": Interviews with Jack Kilby and Willis Adcock; Robert Noyce, "Microelectronics," *Scientific American*, September 1977, p. 69.

165 "We knew we were doing pretty well . . .": Patrick Haggerty, "A New Challenge," remarks at 25th anniversary observance, transistor radio (Dallas: Texas Instruments, 1980), p. 1.

165 Some Texas Instruments people say: Interviews with Adcock and Terry Merryman.

165 "if that little outfit down in Texas . . .": Haggerty, "A New Challenge," p. 1.

166 Haggerty started talking: Interview with Kilby.

167 "I sort of defined . . .": Ibid.

167 "It was a miserable . . ."; "He's one of these guys . . .": Ibid.

168 "that you're going to find an answer if you think . . .": Interview with Merryman.

169 "The basic rule was . . ."; "That way, all the bits . . .": Ibid.

170 "We were trying . . .": Ibid.

171 "the thing was all laid out . . .": Ibid.

172 "How many housewives . . .": Ernest Braun and Stuart MacDonald, *Revolution in Miniature*, p. 175.

173 the first digital watch: Editors of Electronics, *An Age of Innovation*, p. 398.

174 The story of the microprocessor begins: Robert Noyce and Marcian E. Hoff, "A History of Microprocessor Development at Intel," *IEEE Micro*, February 1981, p. 8 ff; Howard Libes, "The First Ten Years of American Computing," *Byte*, July 1978, p. 64 ff.

175 "If this continued . . .": Noyce and Hoff, p. 8.

176 Accordingly, Busicom told Intel: Ibid., p. 13.

177 "insert intelligence into many products . . .": Ibid.

177 "computer on a chip": Cf. advertisement in *Electronics News*, Nov. 15, 1971.

177 the first patent awarded for a microprocessor: U.S. Patent No. 3,757,306.

177 actually was a computer on a chip: U.S. Patent No. 4,074,351.

178 The introductory price was $200: Libes, p. 68.

179 "Project Breakthrough! . . .": *Popular Electronics,* January 1975.

180 Homebrew Computer Club: *Time,* Jan. 3, 1983.

182 "Clearly, a world with . . .": Interview with Noyce.

Chapter 9: DIM-I

183 a summer day in 1976: *The New York Times,* July 11, 1976, sec. 3, p. 13; interview with Dr. Uta Merzbach of the Smithsonian Institution.

183 had watched sales fall: *The New York Times,* Jan. 3, 1982, sec. 4, p. 9.

184 "Calculator usage is now . . .": Ibid., July 11, 1976, sec. 3, p. 13.

184 "That silent computational . . .": Henry Petroski, "Reflections on a Slide Rule," *Technology Review,* February/March 1981, p. 32.

185 "The absence of a decimal point . . .": Ibid., p. 35.

186 "It has a sort of . . .": Interview with Jack Kilby.

189 TMS 1000C: "Semiconductor Products Master Selection Guide" (Dallas: Texas Instruments, 1982), p. 9.

191 This is a fairly standard clock rate: Eugene McWhorter, "The Small Electronic Calculator," *Scientific American,* March 1976, p. 88.

193 propagation delay: Cf. Jacob Millman, *Microelectronics,* pp. 161, 225.

200 LED and LCD: Eugene W. McWhorter, *Understanding Digital Electronics* (Dallas: Texas Instruments, 1984), pp. 1–7.

205 truth table: Scientific American, *Microelectronics,* pp. 29–30.

Chapter 10: Sunset, Sunrise

210 on the morning of June 27: Interview with Jack Kilby.

210 TO ALL TO WHOM . . .: U.S. Patent Office, Letters Patent No. 3,819,921.

211 Since its first appearance in the 1930s: John McPhee, *The Impact of Electronics on the U.S. Calculator Industry, 1965 to 1974* (Washington, D.C.: U.S. Department of Commerce, 1975), pp. 6–8.

212 "With innovative products . . ."; "an excellent illustration . . .": Ibid., pp. 3, 1.

212 TI sued Casio: Interview with Melvin Sharp, patent counsel, TI.

212 "Since 1974, the situation has once again . . .": *The U.S. Calculator Industry Since 1974* (Washington, D.C.: U.S. Department of Commerce, 1977), p. 1.

213 Japanese made about 45 percent: Ibid., p. 7.

213 By the early 1980s: Interview with John McPhee, Office of Business Research and Analysis, U.S. Department of Commerce.

213 "This pattern of market activity . . .": *The U.S. Calculator Industry Since 1974*, p. 1.

214 television history: G.W.A. Dummer, *Electronic Inventions and Discoveries*, 3rd ed., p. 27; Bingley, "A Half-Century of Television Reception," *IRE Proceedings*, May 1962, pp. 799–802; S. Handel, *The Electronic Revolution*, pp. 68–72, 129 ff.

214 the term "television": Elizabeth Antébi, *The Electronic Epoch*, p. 149.

215 Zworykin; Farnsworth: Handel, pp. 68–72.

215 6,000 television sets . . . two years later, 7 million: Jack Baranson, *The Japanese Challenge to U.S. Industry* (Lexington, Mass.: Lexington Books, 1981), p. 75.

216 U.S. television manufacturers developed a marketing: U.S. Congress, Office of Technology Assessment, *U.S. Industrial Competitiveness* (Washington, D.C.: U.S. Congress, 1981), p. 78. Hereafter called OTA Study.

216 purchased the rights to more than 400 patents: Baranson, p. 38.

216 Japanese television firms approached: OTA Study, pp. 78–79; National Science Foundation, *The Five-Year Outlook for Science and Technology*, vol. 2 (Washington, D.C.: U.S. Government Printing Office, 1981), pp. 409–10.

217 The results: Baranson, pp. 75–78; *Five-Year Outlook*, pp. 409–11.

217 But the color TV market turned into: Baranson, p. 77.

217 Admiral; Motorola: Ibid., table 5–1, p. 77.

217 some Japanese competitors were "dumping"; "orderly marketing agreement": Ibid., p. 82; OTA Study, pp. 114–16.

218 to a single chart: *The U.S. Consumer Electronic Industry and Foreign Competition* (Washington, D.C.: U.S. Department of Commerce, 1980), p. 2.

219 between 1964 and 1970, royalty payments: John Tilton, *International Diffusion of Technology* (Washington, D.C.: Brookings Institution, 1971), p. 166.

219 For some U.S. firms, particularly Fairchild: Finan, *The International Transfer of Semiconductor Technology Through U.S.-Based Firms* (National Bureau of Economic Research, 1975), p. 47; Interview with Roger Borovoy, Intel Corporation.

219 "American firms have generally . . .": Tilton, p. 166.

220 TI spurned royalty payments . . . the Dallas firm was: Ibid., pp. 146–47; *The Washington Post*, Mar. 23, 1980, p. E2.

220 "vision," for the 1980s: U.S. Congress, Joint Economic Committee, *International Competition in Advanced Industrial Sectors: Trade and Development in the Semiconductor Industry* (Washington, D.C.: U.S. Congress, 1982), p. 56.

221 In the 1950s, for example . . . Japan's foray into: Interview with Kobayashi; OTA Study, pp. 78–92.

221 Just as MITI planned it: OTA Study, pp. 79–94.

222 By 1980 the Japanese had 42 percent . . . By 1983 they: *The Wall Street Journal*, Nov. 17, 1981, p. 1.

223 "The television people woke up . . .": *The New York Times*, Jan. 18, 1978, p. D1.

224 "I just don't want to pretend . . .": Ibid., Apr. 4, 1982, sec. F, p. 14.

224 The dissenters pointed out: *The Washington Post*, Mar. 23, 1980, p. E2; *Inquiry*, November 1983, p. 12.

225 "The Anderson Bombshell": *The Rosen Electronics Letter*, Mar. 31, 1980, p. 3.

225 Anderson paper: Printed in U.S. House of Representatives, Committee on Ways and Means, *Market Conditions and International Trade in Semiconductors*, no. 96–60, Apr. 28, 1980, pp. 36–41.

227 "U.S. Microelectronics Firms . . .": *The Wall Street Journal*, Nov. 17, 1981, p. 20.

228 "Well, I'm not doing anything . . .": Interview with W. Edwards Deming.

229 "14 points": Deming, *Quality, Productivity, and Competitive Position*, pp. 11–17.

229 "It's so simple . . ."; "By describing statistically . . .": Deming, lecture at Falmouth, Massachusetts, Mar. 25, 1983.

230 "To save 15 cents per spool . . ."; ". . . before it gets to your inspectors?": Interview with Deming.

231 "The total cost to produce and dispose of . . .": Deming, *Quality, Productivity, and Competitive Position*, p. 21.

231 "Brilliant applications burned, sputtered . . .": Ibid., p. 101.

232 "I predicted . . ."; "Once you convince . . .": Interview with Deming.

233 Third Wave: King, "Introducing the Deming Approach to Management Growth Opportunity Alliance," Lawrence, Massachusetts, 1983, p. 1.

234 "The way to compete . . .": Interview with Deming.

235 began to sound very much: "The International Microelectronic Challenge," Semiconductor Industry Association, May 1982, pp. 21–32, 41.

236 "Many of the Japanese practices are . . .": Field hearing on international policy, U.S. House Committee on Banking, Finance, and Urban Affairs, Aug. 18, 1983, p. 21 (stenographic transcript).

237 "Price wasn't much . . .": Robert Noyce, in *Fortune*, June 20, 1988, p. 58.

239 "mothers who say proudly . . .": *Fortune*, June 5, 1989, p. 248.

239 broader problem of American "decline": James Fallows, *Looking at the Sun* (New York: Pantheon, 1994), chap. 1.

240 American industry had caught up: *The Washington Post*, Nov. 20, 1992, p. A1.

Chapter 11: The Patriarchs

241 "I just sort of drifted into it": Interview with Robert Noyce.

242 "Bob was everybody's choice": Interview with Gordon Moore.

242 "One of the real problems . . . for what can be done": Interview with Noyce.

243 when he and his seven cofounders sold: Gene Bylinsky, *The Innovation Millionaires* (New York: Scribner, 1976), p. 64.

243 "comfortable with risk"; "intuitive gut feel": Interview with Noyce.

245 "Intel became . . .": *The Wall Street Journal*, Feb. 4, 1983, p. 21.

245 In the year 2000: Intel Corporation, *Annual Report, 2000.*

245 at an opening price of: *The Wall Street Journal*, Oct. 14, 1971, p. 25.

246 His wife asked him once: Field hearing on international policy, Subcommittee on Trade, U.S. House Committee on Banking, Finance, and Urban Affairs, Aug. 18, 1983, p. 26.

247 "He had one helluva life": *The Wall Street Journal*, Oct. 17, 2000, p. A22.

248 "scientist-cum-charmer": Jerry Sanders, quoted in ibid., Feb. 4, 1983, p. 1.

250 "There is a basic . . .": Jack Kilby, "The Inventor's View," *Chemtech*, March 1979, p. 66.

251 "There's a certain amount . . .": Interview with Jack Kilby.

251 "No, it was pretty damn close . . .": Ibid.

251 "Basically, engineers are hired . . .": *CBS Morning News*, Oct. 24, 1983.

253 "There are a large number . . .": *Chemtech*, March 1979, p. 66.

254 "I would say Jack's . . .": Interview with Willis Adcock.

257 "financial genie": *Time*, Jan. 23, 1984, p. 48.

258 "The Chip That Jack Built . . .": www.ti.com/corp/docs/kilbyctr/jackstclair.shtml

260 Sawyer-Kilby interview; "Coming up in a moment . . .": *CBS Morning News*, Oct. 4, 1983 (transcript in author's files).

260 tried to interview: Ibid.

263 "Why? Because he was a businessman . . .": *The Wall Street Journal*, Oct. 17, 2000, p. A22.

263 The official announcement: www.nobel.se/physics/laureates/2000/illpres/kilby.html

264 "I'm sorry he's not still alive . . .": *The Dallas Morning News*, Oct. 11, 2000.

264 "He looked like everyone's grandfather . . .": Ibid.

265 Kilby's Nobel lecture: www.nobel.se/physics/laureates/2000/kilbyvideo.html

266 "A Kansas ice storm . . .": *Topeka Beacon-Journal*, Dec. 31, 2000, p. 1.

INDEX